Controllable Synthesis, Structure and Property Modulation and Device Application of One-Dimensional Nanomaterials

Controllable Synthesis, Structure and Property Modulation and Device Application of One-Dimensional Nanomaterials

Proceedings of the 4th International Conference on
One-Dimensional Nanomaterials (ICON2011)

Beijing, China,   7–9 December, 2011

Editor-in-chief
Yue Zhang
University of Science and Technology Beijing, China

Editors
Yousong Gu
Yunhua Huang
Xiaoqin Yan
Qingliang Liao
University of Science and Technology Beijing, China

World Scientific

NEW JERSEY · LONDON · SINGAPORE · BEIJING · SHANGHAI · HONG KONG · TAIPEI · CHENNAI

Published by

World Scientific Publishing Co. Pte. Ltd.

5 Toh Tuck Link, Singapore 596224

USA office: 27 Warren Street, Suite 401-402, Hackensack, NJ 07601

UK office: 57 Shelton Street, Covent Garden, London WC2H 9HE

British Library Cataloguing-in-Publication Data
A catalogue record for this book is available from the British Library.

CONTROLLABLE SYNTHESIS, STRUCTURE AND PROPERTY MODULATION AND DEVICE APPLICATION OF ONE-DIMENSIONAL NANOMATERIALS
Proceedings of the 4th International Conference on One-Dimensional Nanomaterials (ICON2011)

ISBN-13 978-981-4407-59-5
ISBN-10 981-4407-59-3

Printed in Singapore by World Scientific Printers.

Preface

The 4th International Conference on One-dimensional Nanomaterials (ICON2011) was held successfully from December 7 to 9, 2011 in Beijing, China. This conference provides a platform for communications among researchers in the field of one-dimensional nanomaterials to exchange ideas, present achievements and discuss the future of one-dimensional nanomaterials. It is a great get together for top experts in the world in the field of one dimensional nanomaterial. It can serve the interest of our scientific communities, as well as industries and general public at large.

ICONs were held successfully in Chinese Taipei in 2005, Sweden in 2007 and USA in 2009. It's the first time for ICON to come to mainland China and we are delight to see a great increase in participants. There are 8 keynote speakers, 24 invited speakers and 36 oral presenters and 67 posters.

In recent years, rapid progresses have been made in both fundamental research and technique applications of one-dimensional nanomaterials such as nanowires, nanofibers, nanobelts, nanorods, etc. due to their remarkable physical and chemical properties as well as wide potential in applications. Rational design and controllable production of one-dimensional nanostructures are the basis for optimization the performance of nanoscale devices and systems for application in electronics, optoelectronics and spintronics etc.

The conference focused on the rational synthesis, structure modulation and property optimization, device fabrication, system integration and novel applications of 1D nanomaterials in nanoelectronics, nano-optoelectronics, nanophotonics, nanopiezotronics, biomedical sciences, sensors, energy technology (solar cell, thermoelectric, piezoelectric nanogenerator, energy storage), and environmental sciences.

The proceedings of the conference are divided into 8 sections: (1) Growth mechmics and structure modulation, (2) property modulations, (3) theoretical simuluaitons, (4) photoic devices, (5) bio-sensors, (6) mechanical detectors (7) solar cells (8) nanodamage.

We would like to give our sincere thanks to all of our new and old friends, who have come to Beijing to attend the conference, and special thanks to the authors who contribute their papers to the conference proceedings.

We would like to thank Ph.D students Zi Qin, Xiaohui Zhang, Zheng Zhang, Qi Zhang, Yanguang Zhao, Guangjie Zhang. They spent a lot of time on proof reading the contributed papers, checking and correcting the content, language and format.

This conference was sponsored by the National Natural Science Foundation of China, the Ministry of Education of the People's Republic of China, University of Science and Technology Beijing, FEI Company, and JEOL Ltd, etc.

Yue Zhang
University of Science and Technology Beijing

The 4th International Conference on One-dimensional Nanomaterials

Bejing 7-9 December, 2011

Organizer

Research Center for Nanoscience and Technology,
University of Science and Technology Beijing, China

Sponsors

National Natural Science Foundation of China
Ministry of Education of the People's Republic of China
University of Science and Technology Beijing
FEI Company

Advisory Board (Alphabetical)

Prof. Yoshio Bando	National Institute for Materials Science, Japan
Prof. Shoushan Fan	Tsinghua University, China
Prof. Ruiping Gao	National Natural Science Foundation of China, China
Prof. Runsheng Gao	Ministry of Education of China, China
Prof. Minghong He	National Natural Science Foundation of China, China
Prof. Ming Li	National Natural Science Foundation of China, China
Prof. Charles M. Lieber	Harvard University, USA
Prof. Zhongfan Liu	Peking University, China
Prof. Ke Lu	Institute of Metal Research, CAS, China
Prof. Chong-Yun Park	Sungkyunkwan University, Korea
Prof. Lars Samuelson	Lund University, Sweden
Prof. Sishen Xie	Institute of Physics, CAS, China
Prof. Ningsheng Xu	Sun Yat-sen University, China
Prof. Qikun Xue	Tsinghua University, China
Prof. Jing Zhu	Tsinghua University, China
Prof. Xing Zhu	Peking University, China

Conference Chairs

Prof. Ze Zhang	Zhejiang University, China
Prof. Yue Zhang	University of Science & Technology Beijing, China
Prof. Zhonglin Wang	Georgia Institute of Technology, USA

Organizing Committee (Alphabetical)

Prof. Xuedong Bai	Institute of Physics, CAS, China
Prof. Kexin Chen	National Natural Science Foundation of China, China
Prof. Huiming Cheng	Institute of Metal Research, CAS, China
Prof. Le Si Dang	CNRS and Université de Grenoble, France
Prof. Hongjin Fan	Nanyang Technological University, Singapore
Prof. Lin Guo	Beihang University, China
Prof. Wanlin Guo	Nanjing University of Aeronautics & Astronautics, China
Prof. Xiaodong Han	Beijing University of Technology, China
Prof. Zheng Hu	Nanjing University, China
Prof. Sang-Woo Kim	Sungkyunkwan University, Korea
Prof. Cheol-Jin Lee	Korea University, Korea
Prof. Liwei Lin	University of California at Berkeley, USA
Prof. Yichun Liu	Northeast Normal University, China
Prof. Sanjay Mathur	University of Cologne, Germany
Prof. Guowen Meng	Institute of Solid State Physics, CAS, China
Prof. Lianmao Peng	Peking University, China
Prof. Chen Wang	National Center for Nanoscience& Technology, China
Prof. Dapeng Yu	Peking University, China
Prof. Shuhong Yu	University of Science & Technology of China
Prof. Magnus Willander	Linköping University, Sweden

Contents

GRAPHENE AND GRAPHENE-BASED NANOCOMPOSITES: SYNTHESIS AND SUPERCAPACITOR APPLICATIONS

GONGKAI WANG[†], CHANGSHENG LIU

Key Laboratory for Anisotropy and Texture of Materials of Ministry of Education, Northeastern University, Shenyang, Liaoning 110004, China

XIANG SUN, FENGYUAN LU, HONGTAO SUN, JIE LIAN[*]

Department of Mechanical, Aerospace & Nuclear Engineering, Rensselaer Polytechnic Institute, Troy, New York 12180, USA

Graphene, a single-atomic-thick sheet, consists of sp^2-bonded carbon atoms in hexagonal lattice and possesses excellent physical and chemical properties such as high surface area, conductivity, mechanical strength and lightweight. The two-dimensional geometry of graphene nanosheets is ideal for many applications especially in electrochemical energy storage. However, the large-scale production of graphene materials is still a bottleneck limiting the development of advanced energy storage devices. The production of graphene oxide is one of the most critical restrictions in term of synthesis of graphene by wet chemical methods. We have demonstrated the synthesis of high quality graphene oxide by simple chemical reactions with less exo-therm involved without emission of toxic gases, which is more favorable than conventional methods for commercialized synthesis of graphene materials (Figure 1.). Meanwhile, pseudo-capacitive materials such as Co_3O_4 and $MnSn(OH)_6$ coupled with graphene nanosheets were synthesized by soft chemical methods in order to utilize both advantages from electrical double layer and pseudo-mechanisms. Their electrochemical properties were evaluated and the potential applications used as high performance electrodes for capacitive energy storage were discussed as well.

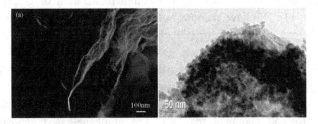

Figure 1. TEM images of graphene nanosheets (a) and $MnSn(OH)_6$/graphene nanocomposite (b)

[†] First author: Tel: 15840345537; Email address: wang.gongkai@gmail.com
[*] Corresponding author: Tel: +1 5182766081; Email address: LIANJ@rpi.edu

1. Introduction

The diverse properties of graphene and graphene-based materials have led to the development of various proof-of-concept devices as well as the demonstration of fundamental science. Graphene, as a unique carbonaceous material, consists of sp^2-bonded carbon atoms in hexagonal lattice [1,2]. The unique two dimensional geometry possesses superhigh surface area (2630 m^2/g) combined with excellent physical and chemical properties (high electrical conductivity and stability under electrochemical conditions), and thus graphene holds tremendous promise in a wide range of applications, especially in energy-storage devices [3-6]. However, the scalable synthesis of graphene that can retain the unique properties at a large extent limits the development of graphene-based materials as well as the potential application in graphene-related downstream industries. Extensive efforts have been focused primarily on the large-scale synthesis of graphene-based materials using various methods which are compatible with industrial processes at a reasonable cost, including chemical exfoliation in/without solvent [7-9]. Large area and high quality graphene can be produced by chemical vapor deposition (CVD) method, showing a promising potential for nanoelectronics. However, the CVD approach does not meet the requirements of using bulk quantities of graphene for achieving desirable volumetric energy and power densities for energy-storage applications. By contrast, graphene nanosheets obtained by controlling chemical exfoliation in/without solvent display greater potentials towards the mass production [2].

Graphene oxide (GO), as the important precursor for the generation of graphene by chemical exfoliation, strongly affects on the structure and quality of graphene. The individual or few layer graphene can be achieved by overcoming the van der Waals forces existing between graphene interlayers. Therefore, controllable synthesis of GO is of scientific interests and meanwhile technological important for driving the commercialization of graphene-based materials [10, 11]. We have demonstrated the synthesis of high quality or high oxidation degree of GO by simple chemical reactions with less exo-therm involved without emission of toxic gases, which is more favorable than the conventional method (the Hummers method) [12] for large-scale synthesis of graphene materials [13]. In this paper, pseudo-capacitive materials such as Co_3O_4 and $MnSn(OH)_6$ were incorporated into as-prepared graphene nanosheets synthesized by soft chemical methods. The graphene-based nanocomposites highlight the concept of combining advantages of both electrical double layer and pseudo-mechanisms for electrochemical energy storage. Their electrochemical properties were evaluated and the potential applications used as high performance electrodes for capacitive energy storage were discussed as well.

2. Experimental

2.1 Synthesis of graphene materials

2.1.1. Controlled synthesis of graphene oxide

We have explored different chemical methods to synthesize high quality GO, as the precursor for large-scale production upon thermal or chemical reductions. Four typical samples are selected, denoted as No-1, No-2, No-3 and No-4, respectively. In brief, the No-1 and No-2 samples were synthesized by first soaking graphite powders in concentrated H_2SO_4 for 12 h followed by adding $KMnO_4$, and stirred for 1 h at room temperature and then heated at 85 °C for 1h (The No-2 sample) or stirred at room temperature for 3 h only without heating (The No-1 sample). The GO No-3 was prepared by the conventional Hummers method for comparison. The No-4 sample was acid treated graphite by H_2SO_4/HNO_3 (3:1) for 4 h at room temperature as a control sample [13].

2.1.2. Synthesis of graphene nanosheets

Graphene nanosheets were produced by thermal exfoliation of as-synthesized GO powders. Generally, the GO (200 mg) was loaded in a crucible in a quartz tube under the argon protective environment. The quartz tube was quickly inserted into the heating zone of a tube furnace (GSL1100X, MTI, USA) which was preheated to 1050 °C and held there for 30 s. The GO can be exfoliated by rapid heating (2000 °C/min) into few layered graphene sheets [14].

2.2 Synthesis of graphene-based nanocomposites

For the exploration of supercapacitor applications, two pseudo-capacitive materials of Co_3O_4 and $MnSn(OH)_6$ coupled with as-prepared graphene nanosheets were prepared. In the case of $MnSn(OH)_6$/graphene nanocomposites [15], two typical samples of monolithic $MnSn(OH)_6$ and $MnSn(OH)_6$/graphene composite were selected, denoted as M and M+G-0-5h, respectively. Briefly, appropriate amount of $MnCl_2 \cdot 4H_2O$ and $Na_2SnO_3 \cdot 3H_2O$ of the molar ratio 1:1 were added into the graphene nanosheets suspension. The mixture was kept at 0 °C for 5 h. The final composite sample was collected by vacuum filtration and dried in air at 80 °C for 12 h. For comparison, a pure $MnSn(OH)_6$ control sample was also prepared followed the same procedure at room temperature for 24 h without the addition of the graphene nanosheets.

For the Co_3O_4/graphene nanocomposites, the sample was synthesized by a surfactant directed growth method as reported previously [16]. Typically, appropriate amount of $CoCl_2$, sodium dodecyl sulfate (SDS), urea, as-prepared graphene and DI water were stirred at 40 °C for 1 h to obtain a transparent solution followed by further maintaining at 80 °C for 6 h. The obtained slurry was filtered then dried at 120 °C overnight. The final products were washed with ethanol and sintered at 400 °C, denoted as Co_3O_4+5G.

2.3 Characterization

The X-ray photoelectron spectroscopy (XPS) was carried out on a PHI 5000 Versa Probe system. Raman spectroscopy was recorded with a Jobin-Yvon HR300 Raman spectrometer equipped with a 532 nm green laser source. The UV-vis spectra were acquired with a VARIAN 6000i UV-vis-IR spectrophotometer. The X-ray diffraction (XRD) was performed on a PANanalytical X-ray diffraction system with a source wavelength of 1.542 Å at room temperature. The morphology and microstructures of the samples were characterized by a field emission scanning electron microscopy (FESEM) (JSM-6335) and a transmission electron microscopy (TEM) (JEOL 2010) operated with 200 KeV electron beam.

The electrochemical properties of the samples were performed on an electrochemical workstation (Ametek, Princeton Applied Research, Versa STAT 4).

3. Results and discussion

The prepared GO samples were analyzed systematically by XRD, XPS, Raman and UV-vis. The correlation among the interlayer d-spacing of GO from XRD, the oxidized carbon percentage from XPS and the absorption wavelength from UV-vis spectra is shown in Figure 2., demonstrating the GO sample with a greater degree of oxidation displays a larger d-spacing, a higher percentage of oxidized carbon and a lower absorption wavelength [13].

Graphene nanosheets were obtained by thermal exfoliation. The optical images of GO and graphene are shown in Figure 3. (a). The GO powder presents yellow brown features, and upon thermal reduction. After the thermal exfoliation, the collected graphene powder shows black color by contrast, demonstrating the GO was reduced from insulator to the semimetal possessing the nature of pristine graphite. Meanwhile, it is worthy of noting that the volume of the collected graphene powder is much larger than that of GO despite the mass of graphene is several times less than that of the original GO, indicating the graphene platelets was achieved by overcoming the van der Waals between GO basal planes through the thermal exfoliation process. The pristine graphite, GO and graphene nanosheets were also evaluated by Raman spectroscopy shown in Figure 3. (b). The Raman spectrum of pristine graphite presents the prominent G band located at 1581 cm^{-1} with a rarely small D band. The G band of the spectrum of GO becomes broaden while the D band becomes prominent, indicating the increase of the size of sp^3 domains. The increased D/G ratio for graphene as compared with that of GO can be ascribed to the reduction of the average size of sp^2 domains and the formation of new graphitic domains. These results are similar to the previous report [17].

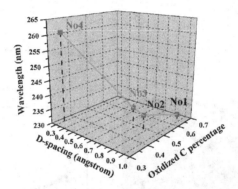

Figure 2. The correlation among d-spacing, the oxidized carbon percentage and the absorption wavelength of the GO samples [13].

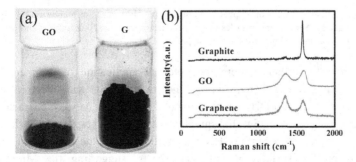

Figure 3. (a) Optical images of GO and graphene. (b) Raman spectra of pristine graphite, GO and graphene.

6

The morphology and microstructure of graphene was characterized by TEM and FESEM shown in Figures 4. (a) and (b). The images indicate that the graphene nanosheets possessing a highly wrinkled surface topology with ultrathin thickness (<5 nm) consistent with the previous observation[18]. The surface area of the thermal exfoliated graphene nanosheets confirmed by N_2 absorption Brunauer-Emmett-Teller (BET) measurement is about 500 m^2/g (Figure 4. (c)), which is comparable with other reports [3, 19]. The pore size of the graphene sample obtained from BET profiles distributes mainly at around 4 nm (Figure 4. (d)), suggesting the graphene nanosheets maintain porous structures.

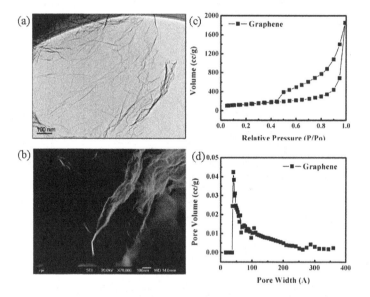

Figure 4. Morphology images of graphene nanosheets. (a) TEM image of graphene [15]. (b) SEM image of graphene. Surface area profiles of graphene nanosheets. (c) N_2 adsorption and desorption isotherms of graphene. (d) Pore size distribution profile of graphene.

In order to further explore the potential application of graphene-based nanocomposites for supercapacitors, the morphology and electrochemical performances of MnSn(OH)$_6$/graphene and Co$_3$O$_4$/graphene were characterized. The TEM images of MnSn(OH)$_6$/graphene and Co$_3$O$_4$/graphene are shown in Figures 5. (a, b). The MnSn(OH)$_6$ nanoparticles with the size of 20 nm are dispersed on the surface of graphene sheets, suggesting the MnSn(OH)$_6$ nanoparticles can be well utilized attributed to the good conductive connection with graphene used as electrode materials for energy storages [15]. Similarly, the particle size of the as-synthesized Co_3O_4 is about 20 nm. All of the particles are attached to the surface of graphene sheets.

The electrochemical performances for supercapacitors of the selected samples were tested by cyclic voltammograms (CV) and galvanostatic charge/discharge as shown in Figures 6. (a-d). Figure 6. (a) shows a distorted shape of CV curves of pure $MnSn(OH)_6$, the Figure 6. (b) shows the relative rectangle CV curves of sample M+G-0-5h by contrast, proving the charge transport of the electrodes were improved by integrating with graphene nanosheets. The calculated specific capacitance is 59.4 F/g based on the mass of $MnSn(OH)_6$ nanoparticles [15]. Figure 6. (c) presents the CV curves of samples Co_3O_4+5G. Two pairs of redox reaction peaks appeared during the test. No obvious changes under various scan rates were observed in regards to the shape of the CV curves, indicating the high charge propagation efficiency within the electrode and low charge/diffusion resistance. The calculated specific capacitance of Co_3O_4+5G is about 300 F/g at the scan rate of 5 mV/s. The electrochemical cyclic stability was recorded by galvanostatic charge/discharge. Both graphene-based nanocomposites displayed an excellent cyclic stability without degradations after 500 cycles (Figure 6. (d)), demonstrating the great potential for supercapacitors application by incorporating graphene nanosheets with pseudo-capacitive materials.

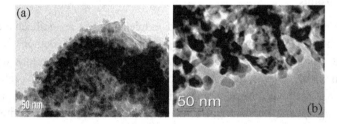

Figure 5. Morphology images of graphene-based nanocomposites. (a) TEM image of $MnSn(OH)_6$/graphene. (b) TEM image of Co_3O_4/graphene.

8

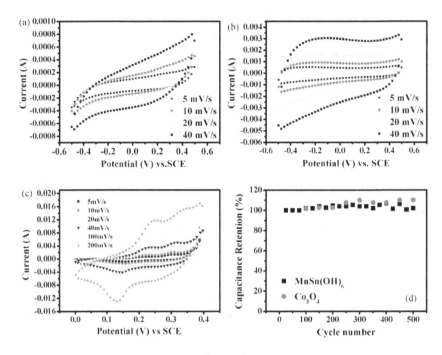

Figure 6. Electrochemical properties of the typical samples. (a) CV curves of the pure MnSn(OH)₆ [15]. (b) CV curves of the M+G-0-5h [15]. (c) CV curves of the Co₃O₄+5G. (d) Average capacitance retention of the M+G-0-5h and Co₃O₄+5G.

4. Conclusions

Large-scale production of graphene and graphene-based are still challenging, limiting the potential applications of graphene. We have demonstrated the controlled synthesis of high quality GO by simple chemical reactions with less exo-therm involved without emission of toxic gases is more favorable than conventional methods toward commercialization. Using $MnSn(OH)_6$/graphene and Co_3O_4/graphene nanocomposites as examples, the potential of integrating graphene nanosheets with pseudo-capacitive nanomaterials for enhancing electrochemical performances was demonstrated. The graphene-based nanocomposites hold tremendous potential for energy-storage applications, especially in the field of supercapacitors.

Acknowledgments

The synthesis of GO and the preparation of graphene upon thermal exfoliation were financially supported by a U.S. DOD-Defense Threat Reduction Agency (DTRA) under the award HDTRA1-10-1-0002. The authors also acknowledge the Graduate Research Innovation Project of Ministry of Education of China (N100602003) and the State Scholarship Fund of China Scholarship Council.

References

1. A. K. Geim and K. S. Novoselov, *Nat. Mater.* **6**, 183 (2007).
2. G. Eda and M. Chhowalla, *Acs Nano* **5**, 4265 (2011).
3. M. D. Stoller, S. J. Park, Y. W. Zhu, J. H. An, and R. S. Ruoff, *Nano Lett.* **8**, 3498 (2008).
4. S. Vivekchand, C. Rout, K. Subrahmanyam, A. Govindaraj, and C. Rao, *J. Chem. Sci.* **120**, 9 (2008).
5. H. Wang, H. S. Casalongue, Y. Liang, and H. Dai, *J. Am. Chem. Soc.* **132**, 7472 (2010).
6. X. Yang, J. Zhu, L. Qiu, and D. Li, *Adv. Mater.* **23**, 2833 (2011).
7. Y. W. Zhu, S. Murali, W. W. Cai, X. S. Li, J. W. Suk, J. R. Potts, and R. S. Ruoff, *Adv. Mater.* **22**, 5226 (2010).
8. G. Q. Ning, Z. J. Fan, G. Wang, J. S. Gao, W. Z. Qian, and F. Wei, *Chem. Commun.* **47**, 5976 (2011).
9. A. K. Geim, *Science* **324**, 1530 (2009).
10. D. R. Dreyer, S. Park, C. W. Bielawski, and R. S. Ruoff, *Chem. Soc. Rev.* **39**, 228 (2010).
11. R. Ruoff, *Nat. Nanotechnol.* **3**, 10 (2008).
12. W. S. Hummers and R. E. Offeman, *J. Am. Chem. Soc.* **80**, 1339 (1958).
13. G. K. Wang, X. Sun, C. S. Liu, and J. Lian, *Appl. Phys. Lett.* **99**, 053114 (2011).
14. J. Rafiee, M. A. Rafiee, Z.-Z. Yu, and N. Koratkar, *Adv. Mater.* **22**, 2151 (2010).
15. G. Wang, X. Sun, F. Lu, Q. Yu, C. Liu, and J. Lian, *J. Solid State Chem.* **185**, 172 (2012).
16. X. Sun, G. K. Wang, J. Y. Hwang, and J. Lian, *J. Mater. Chem.* **21**, 16581 (2011).
17. S. Stankovich, D. A. Dikin, R. D. Piner, K. A. Kohlhaas, A. Kleinhammes, Y. Jia, Y. Wu, S. T. Nguyen, and R. S. Ruoff, *Carbon* **45**, 1558 (2007).
18. D. C. Marcano, D. V. Kosynkin, J. M. Berlin, A. Sinitskii, Z. Z. Sun, A. Slesarev, L. B. Alemany, W. Lu, and J. M. Tour, *Acs Nano* **4**, 4806 (2010).
19. Y. Wang, Z. Q. Shi, Y. Huang, Y. F. Ma, C. Y. Wang, M. M. Chen, and Y. S. Chen, *J. Phys. Chem. C* **113**, 13103 (2009).

PROGRESS IN THE HYDROTHERMAL FORMATION OF DISPERSIVE NANO-PARTICLES AND WHISKERS[*]

XIANG LAN[†]

Department of Chemical Engineering, Tsinghua University, Beijing 100084, China

Ultra-fine powders and one-dimensional (1 D) whiskers have attracted much attention owing to their unique structures, fantastic properties and various applications. Hydrothermal technology has been emerged as one of the desirable methods for the synthesis of nano-scale particles and whiskers owing to its unique advantages as the uniformity of the reaction system, the easy control of the morphology and composition of the products, and the low energy consumption, etc. This article presents a comprehensive overview of the progress in the hydrothermal formation of the dispersive nano-particles $(Mg(OH)_2$, $MoNi/\gamma$-Al_2O_3, NiO, MgO, ZnO, SnO_2, and Al_2O_3) and whiskers $(5Mg(OH)_2 \cdot MgSO_4 \cdot 3H_2O$ (513MOS), $Mg(OH)_2$, and $Mg_2B_2O_5$, $AlOOH$, ZnO, and $CaSO_4$), with the aim of promoting the effective utilization of the rich and cheap mineral resources.

1. Introduction

There is a growing interest in the production and the dispersion of ultra-fine powders. A reduction in particle size to the nanometer scale results in various special properties such as the quantum size effects, the high surface area and the lower sintering temperature, etc., which is the base to obtain fine grain size ceramics with advanced properties. Techniques for the nano-scale particle preparation have been developed in the last years and can be divided into three major classes: chemical, mechano-chemical and thermophysical methods. The chemical routes especially the hydrothermal routes are widely used to produce powder products owing to the relative simplicity and the low energy consumption. In the last ten years our group have focused on the hydrothermal formation of dispersive nano-particles such as $Mg(OH)_2$, $MoNi/\gamma$-Al_2O_3, NiO, MgO, ZnO, SnO_2, and Al_2O_3, etc.

As one of the most thriving research fields in material research, one-dimensional whiskers have attracted much attention owing to their unique

[*] This work is supported by the National Science Foundation of China (50174032, 50574051, 50874066)

[†] Work partially supported by grant 2-4570.5 of the Swiss National Science Foundation.

structures, fantastic properties and various applications. Hydrothermal technology has been emerged as one of the desirable methods for the synthesis of whiskers owing to its advantages (such as the uniformity of the reaction system, the easy control of the morphology and composition of the products, and the lower operation temperature, etc.) over other methods as chemical vapor deposition (CVD) or molten salt synthesis (MSS). Using the dispersive nano-particles as the precursors, we have developed some facile hydrothermal ways to synthesis whiskers as the magnesium-bearing whiskers (513MOS, $Mg(OH)_2$, and $Mg_2B_2O_5$), AlOOH, ZnO, and $CaSO_4$.

2. Hydrothermal formation of dispersive nano-particles [1-11]

Magnesium hydroxide is an effective flame retardant, by virtue of its high decomposition temperature, which can be used in a wider range of thermoplastics than aluminum hydroxide. As an ecological and environment-friendly flame retardant, magnesium hydroxide have an important advantage in that it acts simultaneously, both as a flame retardant and as a smoke suppressant, with low or zero evolution of toxic or hazardous by-products. But $Mg(OH)_2$ formed at room temperature are easy to be agglomerated with each other and also show un-regular shape. Treating the agglomerated $Mg(OH)_2$ in hydrothermal solutions containing NaOH or minor amount of $CaCl_2$, the selective adsorption of the ions in $Mg(OH)_2$ planes at elevated temperatures can accelerate the growth of the polar (110) plane and inhibit the growth of stable (001) plane, leading to the formation of dispersive $Mg(OH)_2$ particles with regular hexagonal sheets. Based on the basic research, we've established a semi-scale industrial plant in Qinghai salt lake, designing a recycling pipe-line reactor to carry out the hydrothermal reaction. The use of the pipe-line reactor has the following advantages: sample structure, easy to be manufactured; safe, stable and can tolerate high temperature; cheap, easy to be scaled up.

The hydrothermal modification methods are also used to synthesis dispersive nano-catalysts and particles as $MoNi/\gamma\text{-}Al_2O_{3(core)}$, NiO, ZnO and γ-Al_2O_3. The dissolution-precipitation of the agglomerated poor crystalline precursors at moderate hydrothermal conditions improved the crystallinity of the nano-particles, which favored the dispersion of the nano-catalysts and particles. For example, the hydrothermal modification of γ-Al_2O_3 at 140 °C for 2 hours converted the γ-Al_2O_3 nano-agglomerates to AlOOH nano-plates composed of AlOOH nano-particles. The modified AlOOH showed higher pore volume, BET and adsorption sites for Mo and Ni, which acted as the suitable support for the formation of dispersive nano-catalyst of $MoNi/\gamma\text{-}Al_2O_3$. As shown in Fig. 2c, the

12

bigger XPS peak for Ni after hydrothermal modification confirmed the higher dispersion since in all of the experiments, the adsorption amount of Ni on Al_2O_3 or AlOOH were kept as the same level.

O: Mg : ○ :O : ○ :H : ○ :Ca

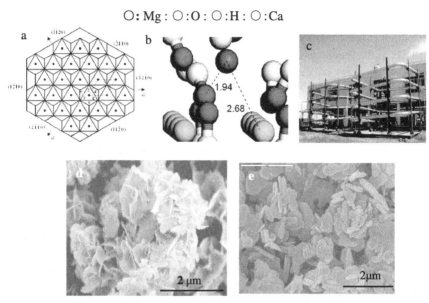

Figure 1. Hydrothermal formation of dispersive $Mg(OH)_2$; (a) structure of $Mg(OH)_2$, (b) adsorption of Ca^{2+} on (001) plane, (c) pipe line reactor (1000 t/year), (d) $Mg(OH)_2$ before hydrothermal modification, (e) $Mg(OH)_2$ after hydrothermal modification.

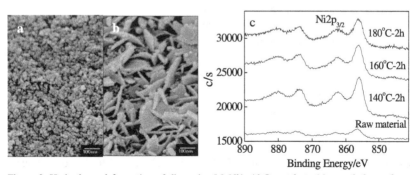

Figure 2. Hydrothermal formation of dispersive MoNi/γ-Al_2O_3 catalyst; (a) morphology of nano γ-Al_2O_3, (b) morphology of AlOOH formed at 140 °C for 2 hours, (c) XPS patterns of Ni at different hydrothermal conditions.

3. Hydrothermal formation of whiskers

Hydrothermal formation of whiskers is an interesting research topic and the controllable growth of the crystals along one direction at elevated temperature using the aqueous solutions as the growth media is the technical key point. With the aim of the potential commercial manufacture of the advanced re-enforcing materials from the rich and cheap mineral resources in China, we have done the work on the hydrothermal formation of the whiskers containing Mg, Al, Zn and Ca, etc.

3.1. *Hydrothermal formation of Mg-bearing whiskers [12-25]*

Using the dispersive $Mg(OH)_2$ nano-particles as the precursors and treating them in hydrothermal solutions containing SO_4^{2-}, BO_3^{2-}, etc., we can form various Mg-bearing whiskers as the $5Mg(OH)_2 \cdot MgSO_4 \cdot 3H_2O$, $Mg(OH)_2$, and $Mg_2B_2O_5$.

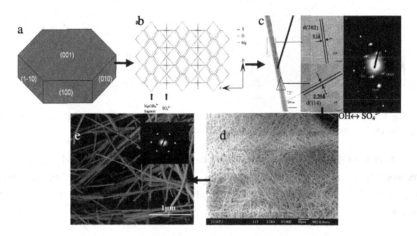

Figure 3. Hydrothermal formation of 513MOS and $Mg(OH)_2$ whiskers; (a) adsorption of SO_4^{2-} on $Mg(OH)_2$, (b) structure of 513MOS, (c-d) morphologies of 513MOS, (e)$Mg(OH)_2$ whiskers.

For example, in the case of the formation of $Mg(OH)_2$ whiskers, it's difficult to form $Mg(OH)_2$ whiskers directly since $Mg(OH)_2$ is a plane structure with an un-regular plate inherent shape. Molecular simulation showed that the hydrothermal shape of $Mg(OH)_2$ is a regular plane considering the influence of temperature and water. $Mg(OH)_2 \cdot MgSO_4 \cdot 3H_2O$ whiskers can be formed by the selective adsorption of SO_4^{2-} on (110) plane of $Mg(OH)_2$ at hydrothermal condition, then converted to $Mg(OH)_2$ whiskers in hydrothermal NaOH solution.

14

The length of the whiskers can be adjusted by adding the complex agents or the surface modification agents.

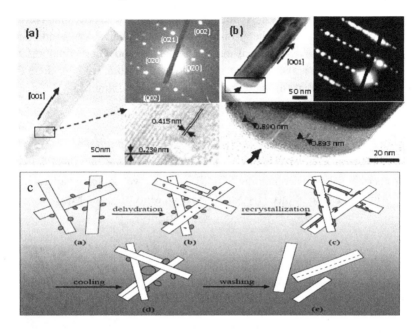

Figure 4. Crystallization of $Mg_2B_2O_5$ in NaCl Flux; (a-b) morphology of $Mg_2B_2O_5$ formed in the absence (a) and presence (b) of NaCl, (c) crystallization process.

MgBO$_2$(OH) whiskers can be formed by treating the nano-Mg(OH)$_2$ in hydrothermal solutions containing BO_3^{3-}. The sintering of the MgBO$_2$(OH) whiskers in NaCl flux led to the formation of $Mg_2B_2O_5$ whiskers with little pores.

3.2. Hydrothermal formation of ZnO whiskers [26-28]

The formation of ZnO with various shapes, esp. the one-dimensional shape was one of the research focuses in the last ten years. Prof. Z.L. Wang in Georgia Institute of Technology has done a lot of excellent research on the formation of various 1D ZnO nano-structures via mainly the CVD route. Prof. Adschiri in Tohoku University has produced the dispersive ZnO nano-particles by the supercritical-hydrothermal method.

Now we are doing the work of hydrothermal formation of ZnO whiskers, using the Mg-bearing ZnSO$_4$, an byproduct in the Zn hydrometallurgical process, as the major raw material. At first, we developed a method to separate Mg^{2+} from $ZnSO_4$, the purified solution was then used to produce Zn-bearing nano-precursors. Nano-ZnO and ZnO whiskers can be formed by treating the Zn-bearing nano-precursors in air and in moderate hydrothermal condition, separately. By careful design of the process, we are trying to establish a clean recycling technology with little produce of waste water or by-products.

3.3. Hydrothermal formation of AlOOH whiskers [29-34]

We have developed a method to synthesis AlOOH whiskers via the hydrothermal treatment of amorphous $Al_2O_3 \cdot xH_2O$ precursor in acidic solutions containing SO_4^{2-} ion. It was found the preferential adsorption of SO_4^{2-} on (010) plane of AlOOH inhibited the growth of the plane, leading to the growth of AlOOH whiskers [100] direction.

Figure 5. Morphology of nano-ZnO (a) and ZnO whiskers (b)

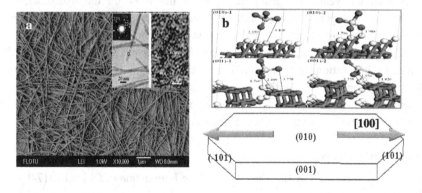

Figure 6. Hydrothermal formation of AlOOH whiskers; (a) morphology of AlOOH whiskers, (b) adsorption of SO_4^{2-} on (010) and (001) planes

3.4. *Hydrothermal formation of CaSO4 whiskers [34-35]*

$CaSO_4 \cdot 2H_2O$ whiskers with a length of 200-1000μm and a diameter of 5-50 μm can be formed at moderate hydrothermal condition with a temperature lower than 160°C. Our work showed that many Ca-bearing ores or the industrial by-products can be used as the raw materials, such as $CaCO_3$, $CaCl_2$, natural gypsum, desulfurization gypsum produced in the metallurgical or power plants, the phosphogypsum produced in the P-fertilizer plants, etc. The gypsum precursors with poor crystallinity, small particle size and less impurities were favorable for the formation of the calcium sulfate whiskers with high aspect ratio.

Figure 7. Morphology of the desulfurization (a) and the $CaSO_4$ whiskers (b).

4. Conclusions

In the last ten years our group has done some work on the hydrothermal formation of dispersive nano-particles and whiskers, with the aim of the effective utilization of the rich and cheap inorganic mineral resources. How to develop the moderate, economic and clean hydrothermal technologies both of academic innovation and industrial application is still an interesting and challenging topic for us. With the continues efforts on hydrothermal fields, we believe that more and more new technologies will be occurred in future and some of them will become one part of the modern industries.

Acknowledgements

The work is supported by the National Science Foundation of China (50174032, 50574051, 50874066).

References

1. L. Xiang, Y. C. Jin, Y. Jin, *Trans. Nonferrous Met. Soc. China* **14**, 370 (2004).
2. H. J. Wu, L. Xiang, Y. C. Jin, Y. Jin, *J. Inorg. Mater.* **19**, 1181 (2004).
3. Z. H. Chen, L. Xiang,Y. C. Zhang, R. Y. Lin, Chin. *J. Inorg. Chem.* **22**, 1062 (2006).
4. Q.L. Wu, L. Xiang, Y. Jin, *Powder Technol.* **162**, 100 (2006).
5. Q. Wang, L. Xiang, Y.C. Zhang and Y. Jin, *J. Mater. Sci.* **43**, 2387 (2008).
6. L. Xiang, Y.L. Gong, J.C. Li, and Z.W. Wang, *Appl. Surf. Sci.* **239**, 94 (2004).
7. J. C. Li, L. Xiang, F. Xu, and Z. W. Wang, *Appl. Surf. Sci.* **253**, 766 (2006).
8. J. C. Li, L. Xiang, F. Xu, Z. W. Wang, and J. Cheng, *Appl. Surf. Sci.* **254**, 2077 (2008)
9. L. Xiang, X.Y. Deng, Y. Jin, *Scripta Mater.* **47**, 219 (2002).
10. H.B. Liu, L. Xiang, Y. Jin, *Cryst. Growth Des.* **6**, 283 (2006).
11. L. Xiang, Y.P. Yin, Y. Jin, *Mater. Lett.* **59**, 2223 (2005).
12. L. Xiang, F. Liu, J. Li and Y. Jin, *Mater. Chem. Phys.* **87**, 424 (2004).
13. J. Li, L. Xiang and Y. Jin, *J. Mater. Sci.* **41**, 1345 (2006).
14. X.T. Sun and L. Xiang, *Cryst. Res. Technol.* **43**, 479 (2008).
15. X. T. Sun and L. Xiang, *Mater. Chem. Phys.* **109**, 381 (2008).
16. X.T. Sun, W.T. Shi, L. Xiang and W.C. Zhu, *Nanoscale Res. Lett.* **3**, 386 (2008).
17. X. T. Sun, L. Xiang, W. C. Zhu and Q. Liu, *Cryst. Res. Technol.* **43**, 1057 (2008).
18. W. C. Zhu, L. Xiang, T. B. He and S. L. Zhu, *Chem. Lett.* **35**, 11 (2006).
19. W.C. Zhu, L. Xiang, X. Zhang and S. L. Zhu, *Mater. Res. Innovations* **11**, 188 (2007).
20. W.C. Zhu, L. Xiang, Q. Zhang, X. Y. Zhang, L. Hu, and S. L. Zhu, *J. Cryst. Growth* **310**, 4262 (2008).
21. W.C. Zhu, Q. Zhang, L. Xiang, F. Wei, X. L. Piao and S. L. Zhu, *Cryst. Growth Des.* **8**, 2938 (2008).
22. W.C. Zhu, X. Y. Zhang, L. Xiang and S. L. Zhu, *Nanoscale Res. Lett.* **4**, 724 (2009).
23. W.C. Zhu, S. L. Zhu and L. Xiang, *CrystEngComm* **11**, 1910 (2009).
24. W.C. Zhu, G. D. Li, Q. Zhang, L. Xiang and S. L. Zhu, *Powder Technol.* **203**, 265 (2009).
25. W.C. Zhu, Q. Zhang, L. Xiang and S. L. Zhu, *CrystEngComm*, in press, 2010.

26. W. T. Shi, G. Gao and L. Xiang, Trans. *Nonferrous Met. Soc. China* **20**, 1049 (2010).

27. G. Gao, W. T. Shi, and L. Xiang, *J. Electrochem. Soc.* **156**, 155 (2009).

28. L.Z. Zhang and L. Xiang, *Res. Chem. Intermed.* **37**, 281 (2011).

29. T. B. He, L. Xiang and S. L. Zhu, *Langmuir* **24**, 8284 (2008).

30. T. B. He, L. Xiang, W.C. Zhu and S.L. Zhu, *Mater. Lett.* **62**, 2939 (2008).

31. T. B. He, L. Xiang, W.C. Zhu and S.L. Zhu, *CrystEngComm* **11**, 1338 (2009).

32. B.H. Hao, K.M. Fang, L. Xiang and Q. Liu, *Intern. J. Miner. Metall. Mater.* **17**, 376 (2010).

33. M. J. Chen and L. Xiang, *Nano Biomed. Eng.* **2**, 121 (2010).

34. M. J. Chen and L. Xiang, *AIP Conf. Proc.* **1251**, 324 (2010).

35. K. B. Luo, C.M. Li, L. Xiang, H.P. Li, and P. Ning, *Particuology* **8**, 240 (2010).

36. K. B. Luo, C. M. Li, L. Xiang, H. P. Li, and P. Ning, *AIP Conf. Proc.* **1251**, 296 (2010).

THE GROWTH OF AL-DOPED ZNO NANOPLATE ARRAYS INFLUENCED BY SOLUTION CONCENTRATION[†]

LIDAN TANG, BING WANG

Department of Materials science and engineering, Liaoning University of technology, Jinzhou, 121001, China

Al-doped ZnO nanoplate arrays had been successfully prepared by using hydrothermal methods. SEM and XRD analysis showed that all of Al-doped ZnO are plate-like nanstructure and high-quality nanocrystal with wurtzite structure. The diameter and thickness of nanoplate are about 400nm and 15nm, respectively when solution concentration is 0.035mol/l. At last the formation mechanism of Al-doped ZnO nanoplate arrays was studied in detail.

1. Introduction

Zinc oxide has attracted much attention as a promising material for many applied devices such as transparent conductor, light emitting diodes, ultraviolet devices, solar cells and piezoelectric devices [1-5], because of its wide band gap and large exciton binding energy. The one-dimensional ZnO had attracted greatest interest because of better properties than other morphological ZnO materials. Especially, Dye-sensitized solar cells based on one-dimensional ZnO nanostructure have attracted a lot of scientific and technological interest in recent years. Various fabrication techniques had been reported, such as chemical vapor deposition [6], metal–organic chemical vapor deposition [7] and hydrothermal method [8–11]. Also, ultraviolet-irradiation method and extra field technology had been introduced to obtain uniform and high-quality nanomaterials[10,11]. Hydrothermal methods are suitable and economical in getting high-quality ZnO nanostructure. These effects of reaction time [12], reaction temperature [13] and pre-coated seed layer [14] on the ZnO nanorob arrays have been investigated extensively. However, Al-doped ZnO nanoplate arrays influenced by the solution concentration were reported firstly. Here Al-doped ZnO nanoplate arrays had been successfully prepared by using

---

[†] This work is supported by the Research Foundation of Education Bureau of Liaoning Province, China (Grant No.L2011098).

hydrothermal methods. The effects of solution concentration on growth of ZnO nanoplate arrays were studied in detail.

2. Experiment procedures

Well-aligned Al doped ZnO nanoplate arrays had been prepared by using hydrothermal methods. Before growth of Al doped ZnO nanoplate arrays, the pre-coated seed layers were deposited on all the Si substrates. Zinc acetate（$Zn(Ac)_2$ • $2H_2O$, 0.5mol/l）, ethanolamine($H_2NCH_2CH_2OH$, 0.5mol/l) and ethylene glycol monomethylether($C_3H_8O_2$) directly dissolved in deionized water to obtain solutions. And then Si substrates immerged in solutions were pulled by using vertical membrane machine for three times to form pre-coated seed layer. At last the pre-coated substrates were heated at 300 °C for10 minute to remove organic matter.

For growth of Al doped ZnO nanoplate arrays the pre-coated substrates were immersed in the aqueous solution containing Zinc nitrate, Aluminum nitrate and hexamethy lenetetramine. The solution concentrations （Zn^{2+} concentration） are 0.015mol/l, 0.025mol/l, 0.035mol/l and 0.045mol/l respectively. The hydrothermal synthesis was carried out at the appointed temperatures of 90 °C for 4 h in the teflon lined stainless autoclaves. After the reaction, the teflon lined autoclave was cooled naturally to the room temperature, and the substrates with as-grown Al doped ZnO nanoplate arrays were rinsed repeatedly with deionized water to remove the residual reactants.

The morphology of Al doped ZnO nanoplate arrays were characterized by using scanning electron microscopy (SEM) (Zeiss, SUPRA-55). The crystalline structure of obtained Al doped ZnO nanoplate arrays were analyzed by X-ray diffraction (XRD) (D/MAX-RB, CuK).

3. Results and discussion

Fig. 1 presents SEM image of Al doped ZnO nanostructure fabricated by hydrothermal methods with various the solution concentration. When solution concentration is too low (0.015 mol/L), ZnO nanostructre did not appeared on the Si substrates. As the solution concentration increases some ZnO nanostructures is not even, as shown in Fig. 1b. When the solution concentration is 0.035 mol/L ZnO nanostructure appeared evenly on the the Si substrates. However, when the solution concentration is too high, ZnO nanostructure disappeared again. As a result the best solution concentration is 0.035 mol/L. In order to observe ZnO nanostructure of Sample C clearly the products was measured by FE-SEM, the typical FE-SEM images were shown in Fig. 2. ZnO showed uniform nanoplated array. The diameter and thickness of nanoplate is about 400nm and 15nm, respectively.

Sample A : 0.015 mol/L

Sample B: 0.025 mol/L

Sample C: 0.035 mol/L

Sample D: 0.045 mol/L

Fig. 1. SEM image of Al doped ZnO nanostructure fabricated by hydrothermal methods with various the solution concentration

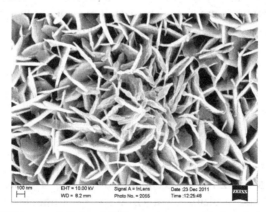

Fig. 2. The FE-SEM image of Al doped ZnO nanostructure with the solution concentration of 0.035mol/L

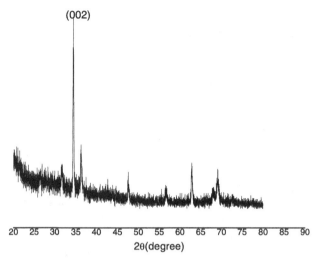

Fig. 3. XRD pattern of ZnO nanoplate arrays synthesized by using solution concentration of 0.035mol/l

Fig. 3 shows XRD pattern of ZnO nanoplate arrays synthesized by using solution concentration of 0.035mol/l. According to the standard XRD patterns, ZnO nanoplate arrays with highly crystallized wurtzite structure are completely formed by hydrothermal methods. Furthermore the intensity of (002) diffraction peaks are much higher than the others, which indicates that ZnO nanoplate are growing along the c-axis and vertically standing on the Si substrate. The XRD results further prove the well orientation of ZnO nanoplate arrays.

Based on the experimental results, the possible formation mechanism of ZnO nanoplate arrays is proposed. The overall reaction for the growth of ZnO nanoplate arrays may be simply formulated as following:

$$(CH_2)_6N_4 + 6H_2O \rightarrow 6HCHO + 4NH_3 \tag{1}$$

$$NH_3 + H_2O \leftrightarrow NH_4^+ + OH^- \tag{2}$$

$$Zn^{2+} + 2OH^- \rightarrow Zn(OH)_2 \tag{3}$$

$$Zn(OH)_2 \rightarrow ZnO + H_2O \tag{4}$$

When the degree of supersaturation exceeds the critical value, reactions 1, 2 and 3 happen. Much Zn $(OH)_2$ colloids exist in the solution, which partly decompose into ZnO nuclei(reaction 4). At the same time the Al doped ZnO seed layer on the Si substrate can selectively adsorb on some specific crystal planes and then restrain the anisotropic growth of ZnO nanoplate, which can drive the ZnO nuclei to form stable and uniform nanoplate in the solution.

4. Summary

Well-aligned ZnO nanorod arrays had been successfully prepared by using hydrothermal methods. SEM and XRD analysis showed that all of Al-doped ZnO are plate-like nanstructure and high-quality nanocrystal with wurtzite structure. the diameter and thickness of nanoplate is about 400nm and 15nm when solution concentration is 0.035mol/l. The formation mechanism of ZnO nanorod arrays had been proposed.

References

1. N. Saito, H. Haneda, T. Sekiguchi, N. Ohashi, I. Sakaguchi, K. Koumoto, *Adv. Mater.* **14**, 418 (2002)
2. T.T. Soki, Y. Hatanaka, D.C. Look, *Appl. Phys. Lett.* **76**, 3257 (2000)
3. A. Umar, Y.B. Hahn, *Cryst. Growth Des.* **8**, 2741 (2008)
4. J.X. Wang, X.W. Sun, A. Wei, Y. Lei, X.P. Cai, C.M. Li, Z.L. Dong, *Appl. Phys. Lett.* **88**, 233106 (2006)
5. D.C. Look, B. Claflin, Y.I. Alivov, S. Park, *J. Phys. Status Solidi* A **201**, 2203 (2004)
6. X.J. Zhua, L.M. Geng, F.Q. Zhang, Y.X. Liu, L.B. Cheng, *J. Power Sources* **189**, 828 (2009)
7. F. Li, Y. Jiang, L. Hu, L. Liua, Z. Li, X. Huang, *J. Alloys and Compounds* **474**, 531 (2009)
8. D. Pradhan, K. Tong, Leung, *Langmuir* **24**, 9707 (2008)
9. L.E. Greene, M. Law, D.H. Tan, M. Montano, J. Goldberger, G. Somorjai, P.D. Yang, *NanoLett.* **5**, 1231 (2005)
10. L. Vayssieres, *Adv. Mater.* **15**, 464 (2003)
11. L.E. Greene, M. Law, J. Goldberger, F. Kim, J.C. Johnson, Y.F. Zhang, R.J. Saykally, P.D. Yang, Angew, *Chem. Int. Ed.* **42**, 3031 (2003)
12. H.Q. Yang, Y.Z. Song, L. Li, J.H. Ma, D.C. Chen, S.L. Mai, H. Zhao, *Cryst. Growth Des.* **8**, 1039 (2008)
13. Z. Qin, Q. L. Liao. Y.H. Huang, L.D. Tang, X.H. Zhang, Y. Zhang, *Mater. Chem. Phy.* **123**, 811 (2010)
14. J. Song, S. Lim, *J. Phys. Chem.* C **111**,596 (2007)

INFLUENCES OF HYDROTHERMAL CONDITIONS ON MORPHOLOGIES OF ZNO NANOWIRE ARRAYS

ZI QIN, YUNHUA HUANG[‡], QINYU WANG

State Key Laboratory for Advanced Metals and Materials, School of Materials Science and Engineering, University of Science and Technology Beijing, 100083 Beijing, P R China

GUANGJIE ZHANG

Key Laboratory of New Energy Materials and Technologies, University of Science and Technology Beijing, 100083 Beijing, P R China

Hydrothermal method is a facile and economical process to synthesize ZnO nanowire arrays. Morphologies, structures, qualities and properties are closely depended on hydrothermal parameters. Influences caused by substrate, duration time, multi-step reaction, and precursor solution concentration have been investigated and discussed in detail. All of these parameters can extremely modulate the morphology of the ZnO nanowire arrays such as the orientation, distribution, diameter and height. The optimal hydrothermal parameters can be selected to synthesize the ZnO nanowire arrays according to the requirement of the devices.

1. Introduction

ZnO nanostructures attract much more interests due to their excellent properties such as wide band gap, good transport properties and facile morphology tailoring. Zinc oxide (ZnO) is a wide band-gap semiconductor, which has a direct wide band gap (3.37 eV) and large exciton binding energy (60 meV). Especially, ZnO nanowire arrays is the most famous one in ZnO nanostructures, which have been used to fabricate lots of devices such as dye-sensitized solar cells [1-5], ultraviolet devices [6,7], biosensors [8] and piezoelectric devices [9]. The optimum performance of each device depends on the morphology and structure of ZnO nanowires. Especially in recent years, ZnO nanowire arrays is considered to be one of the most promising oxide semiconductor materials to instead of the typical TiO₂ used as the photoelectrode, in order to decrease the recombination rate and enhance the DSCs performance. ZnO nanowire arrays

---

[‡] Corresponding author's E-mail: huangyh@mater.ustb.edu.cn.

may significantly enhance the cell performance because of the direct transport pathways for photo excited electrons. So the synthesis of ZnO nanowires is primary task for the devices based on ZnO nanowire arrays. As reported in literatures, ZnO nanowire arrays can synthezied by various methods such as chemical vapor deposition [10], metal-organic chemical vapor deposition [11], pulsed laser deposition [12], electrochemical deposition [13] and hydrothermal method [14-15]. However, hydrothermal method is the more economical process to obtain uniform and high-quality ZnO nanowire arrays. In addition, the morphology of the final production is depended on each reaction parameter in hydrothermal method.

In this work, the morphology influences caused by the growth substrate, reaction duration time, multi-step reaction times, and precursor solution concentration have been investigated and discussed in detail. These entire parameters can obviously affect and modulate the morphology of the ZnO nanowire arrays such as the orientation, distribution, diameter and height. To be worthy of attention, this investigation may play an important role in the further application of ZnO nanowire arrays based devices.

2. Experimental details

Fluorine-doped tin oxide (FTO) glass (1.5 cm×2.0cm, 14ohm/sq, Nippon Sheet Glass, Japan) was used as the growing substrate for ZnO nanostructures. All the substrates were ultrasonically cleaned in acetone and alcohol for several times, and finally rinsed in deionized water and dried in flowing nitrogen gas before further treatment. Pre-treatment colloid solution for the ZnO seed layers contains equivalent zinc acetate dehydrate ($Zn(CH_3COO)_2 \cdot 2H_2O$,) and ethanolamine ($NH_2CH_2CH_2OH$,) were mixed in 2-methoxyethanol ($CH_3OCH_2CH_2OH$), and spin-coating method was used to fabricate ZnO seed layer on the substrates. For the hydrothermal reaction, the precursor solution contained equivalent $Zn(NO_3)_2$ and methenamine, sealed in the teflonlined autoclave under the necessary temperature. After the synthesis, the autoclave was cooled naturally to the room temperature, and the substrates with products were rinsed repeatedly with deionized water to remove the residual reactants. The morphologies of the ZnO nanostructures were characterized by field emission scanning electron microscopy (FE-SEM) (Zeiss, SUPRA-55).

3. Experimental results and discussion

In order to study the effect of the ZnO seed layer for the hydrothermal production, two types of FTO growth substrates were used. One is covered with the ZnO seed layer, and the other has none. The density of the colloid solution to fabricate ZnO seed layers is 0.5M. Both substrates were immerged in the 0.05M hydrothermal solution for 24h at 95°C. The morphologies of the products on the two substrates are shown in Fig. 1. It can be seen that disordered ZnO micro-wires are obtained on the uncovered FTO substrate, while well-aligned ZnO nanowire arrays are yield on the FTO substrate covered with ZnO seed layer. The average diameter of the ZnO micro-wires obtained on bare FTO substrate is about 5μm, and their distribution is disordered, which means that bare FTO is not suitable to fabricate ZnO nanowires. On the contrary, the average diameter of the ZnO naowires is about 200nm, and they present the well-aligned arrays. So it can be concluded that ZnO seed layer is the necessary condition to obtain well-aligned ZnO nanowire arrays.

Figure 1. Morphologies of the pwireuctions: （a）FTO substrate （b）FTO substrate covered with ZnO seed layer.

Except for the substrates, different reaction duration time were used to investigate its influence on the morphology of ZnO nanowire arrays. The single hydrothermal growth was carried out for 4 different duration time (2h, 3h, 5h and 10h), keeping other reaction parameters unchanged. The FTO glass covered with ZnO seed layer was used as the growth substrate. The fabrication process of the ZnO seed layer has been described above. All the substrates were immerged in the 0.05M hydrothermal solution at 95°C. The morphologies of the products depend on different reaction duration time are shown in Fig. 2. All the products are well-aligned ZnO nanowire arrays. After the calculation, the average diameters of the 4 types of ZnO nanowires are 70nm, 90nm, 110nm, and 150nm separately. It obviously indicates that the reaction duration time can affect and modulate the diameter of the ZnO nanowire arrays. With the reaction time prolonged, the diameters obviously increase.

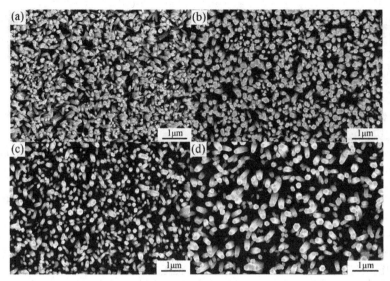

Figure 2. Morphologies of the ZnO nanowire arrays synthesized for different duration time: (a) 2h; (b) 3h; (c) 5h; (d) 10h.

In view of the influence of reaction duration time for single growth, multi-step hydrothermal reaction might has further influence on the ZnO nanowire arrays. Especial to the influence of the height of ZnO nanowire arrays, which has not been studied in the single growth process above. So we select 3 different times (10, 15 and 20) for growth, and the duration time is 12h for each cycle. All the substrates were immerged in the 0.05M hydrothermal solution at 95°C. At the same time, Polyetherimide (PEI) was added in the precursor solution to keep the c-axis growth, and the solution is also refreshed for each time. Fig. 3 (a) to (c) shows the morphologies of the ZnO nanowire arrays yields from the 10, 15 and 20 times hydrothermal reactions. All the products are well-aligned ZnO nanowire arrays. We can also see that the ZnO nanowire arrays grow higher, and the average heights of the three samples are 18μm, 30μm and 40μm, while the diameters of the 3 samples only have a few increases. However, because of the growth of the diameters, the ZnO nanowire arrays form as a compact film at the bottom of the arrays, which means that the large quantities of cycle times is harmful to obtain well ZnO nanowire arrays. But we also can conclude that the moderate multi-step hydrothermal growth can extremely increase the height of the ZnO nanowire arrays.

Figure 3. Morphologies of the ZnO nanowire arrays under multi-steps synthesis: (a)10 times; (b)15 times; (c)20 times.

In the hydrothermal reaction, the concentration of the precursor solution is also an important parameter to the morphology of the production. Therefore, we selected two different solution concentrations to investigate the influence. Precursor solution with 0.005M and 0.05M were used to fabricate ZnO nanowire arrays separately. The reactions were carried out for 4h at 90°C. The morphologies of the productions are shown in Fig. 4. It is obvious that the average diameter of the ZnO nanowires in Fig. 4(a) is smaller than that in Fig. 4(b). The average diameter of the wires got from the 0.005M-solution is 50nm, while the other one got from the 0.05M-solution is 150nm. So it indicates that the concentration of the precursor solution also can modulate the diameter of the ZnO nanowires, and the diameters can extremely increase with the arising of the concentration.

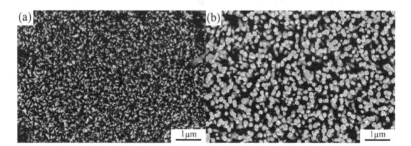

Figure 4. Morphologies of the ZnO nanowire arrays synthesized in different concentration precursor solutions: (a) 0.005M; (b) 0.05M.

4. Conclusions

Hydrothermal reaction parameters present important influences on the morphology of ZnO nanowire arrays. ZnO seed layer is a vital requirement to yield well-aligned ZnO nanowire arrays. The reaction duration time and solution concentration can easily modulate the diameters of ZnO nanowires. Multi-step synthesis can remarkably increase the height of the arrays.

Acknowledgments

This work was supported by the Major Project of International Cooperation and Exchanges (2006DFB51000), NSFC (51172022), NSAF (10876001), the Research Fund of Co-construction Program from Beijing Municipal Commission of Education, the Fundamental Research Funds for the Central Universities.

References

1. M. Law, L.E. Greene, A. Radenovic, T. Kuykendall, J. Liphardt and P.D. Yang, *J Phys Chem B* **110**, 22652 (2006).
2. Q.F. Zhang, C.S. Dandeneau, X.Y. Zhou and G.Z. Cao. *Adv. Mater.* **21**, 1 (2009).
3. Z. Qin, Q.L. Liao, Y.H. Huang, L.D. Tang, X.H. Zhang and Y. Zhang. *Mater. Chem. Phys.* **123**, 811 (2010).
4. Z. Qin, Y.H. Huang, J.J. Qi, Q.L. Liao, Y.P. Yin and Y. Zhang, *Mater. Lett.* **65**, 3506 (2011).
5. Z. Qin, Y.H. Huang, Q.L. Liao, Z. Zhang, X.H. Zhang and Y. Zhang, *Mater. Lett.* **70**, 177 (2012).
6. M.H. Huang, S. Mao, H. Feick, H.Q. Yan, Y.Y. Wu, H. Kind, E. Weber, R. Russo and P.D. Yang, *Science* **292**, 1897 (2001).
7. A. Umar and Y.B. Hahn, *Cryst. Growth Des.* **8**, 2741 (2008).
8. Y. Lei, X.Q. Yan, J. Zhao, X. Liu, Y. Song, N. Luo and Y. Zhang, *Colloids and Surfaces B: Biointerfaces*, **82**, 168 (2011).
9. X.D. Wang, J.H. Song, J. Liu and Z.L. Wang, *Science* **316**, 102 (2007).
10. Y. Dai, Y. Zhang, Q.K. Li and C.W. Nan, *Chem. Phys. Lett.* **358**, 83 (2002).
11. S. Liu and J.J. Wu, *Mater. Res. Soc. Symp. Proc.* **703**, 241 (2002).
12. J.H. Choi, H. Tabata and T. Kawai, *J. Gryst. Growth* **226**, 493 (2001).
13. D. Pradhan and K.T. Leung, *Langmuir* **24**, 9707 (2008).
14. L.E. Greene, M. Law, D.H. Tan, M. Montano, J. Goldberger, G. Somorjai and P.D. Yang, *NanoLett.* **5**, 1231 (2005).
15. L. Vayssieres, *AdV. Mater.* **15**, 464 (2003).
16. L.E. Greene, M. Law, J. Goldberger, F. Kim, J.C. Johnson, Y.F. Zhang, R.J. Saykally and P.D. Yang, *Angew. Chem. Int. Ed.* **42**, 3031 (2003).

MICROSTRUCTURE AND GROWTH MECHANISM OF MN DOPED ZNS NANOBELTS

JUNJIE QI*, QI ZHANG, ZHANQIANG DENG, YUE ZHANG

State Key Laboratory for Advanced Metals and Materials, School of Materials Science and Engineering, University of Science and Technology Beijing, Beijing 100083 China

Mn doped ZnS nanobelts were successfully fabricated by introducing Mn+2 ions in the raw material via a simple thermal evaporation process. X-ray diffraction (XRD), transmission electron microscopy (TEM) and energy dispersive X-ray spectroscopy (EDS) were applied to investigate the structure of the Mn/ZnS fibers. The results indicated that the Mn/ZnS nanobelt is single crystalline with wurtzite structure and grows along [0001] direction. The growth mechanism was investigated indicating that the growth of the Mn doped ZnS nanobelts is governed by the metal-seeded VLS mechanism. The doped ZnS nanobelts have the potential application as promising blocks for the construction of photoluminescent and electroluminescent nanodevices.

1. Introduction

One dimensional semiconductor nanowires/belts have been attracted growing interests in constructing optoelectronic devices and circuits owing to their excellent properties. ZnS is one of the most important II-VI semiconductors compounds that has been widely used as a key material for some devices such as light-emitting diodes [1], flat-panel displays [1], n-window layers for solar cells [2], luminescence devices [3], photonic crystal devices [4]etc. ZnS has a wide bandgap of 3.7~3.9 eV, which is higher than that of ZnO (3.37 eV), GaN (3.39 eV), and 4H-SiC (3.27 eV), and exhibits the luminescence of short wave centered near 337 nm. Compared with the room-temperature thermal energy (25 eV), ZnS exhibits significantly large exciton binding of 40 eV, which makes the exciton's threshold much lower, so ZnS is an excellent candidate for efficient room-temperature exciton short-wave devices[5]. In addition, doped ZnS[6,7] also has good optical properties for application in semiconductor materials. Therefore, it is believed that one-dimensional ZnS nanostructures, especially doped nanostructures, will exhibit characteristic optical properties.

---

* Corresponding author, e-mail: junjieqi@mater.ustb.edu.cn

ZnS doped with manganese Mn exhibits attractive light-emitting properties with increased optically active sites for applications as efficient phosphors [8,9]. Although 1D ZnS nanostructures [10-13] have been fabricated more and more recently while the Mn doped ZnS nanostructures have been reported rarely. A halide-transport chemical vapor deposition (HTCVD) method was reported to produce Mn and other transition-metal doped ZnS 1D nanostructures [14]. But the fabrication temperature of one-dimensional ZnS nanostructures is usually very high.

In this letter, Compared with the traditional thermal evaporation techniques (1150℃), the high-quality Mn doped ZnS nanobelts with high aspect ratio have been synthesized in a large scale by H_2-assisted thermal evaporation method at lower temperature with Au as catalyst, the microstructure and growth mechanism of the product have been analyzed.

2. Experimental section

The ZnS nanobelts were synthesized by hydrogen-assistant thermal evaporation method through the direct reaction of Zn vapor with S vapor in a furnace with a horizontal quartz tube. First, Zn powder, S powder, and $MnCl_2$ with the molar ratio of Zn : S : Mn = 4:8:1 were thoroughly grinded into a mixture before being loaded into a quartz boat. Then, the Si substrate with 6 nm catalyst with the polished side facing powder was fixed upon the boat, and the boat with the mixture was placed at the center of the furnace. A carrier gas of high-purity argon premixed with 3% hydrogen was kept flowing through the tube at a high flow rate (>400 sccm) (standard cubic centimeters per minute) for 1 h. The process was carried out at 900°C for 30 min. After the reaction was completed, a red-brown product was collected from the Si substrate.

The morphologies and structures of the synthesized product were characterized using X-ray diffraction (XRD) (Rigaku DMAX-RB, Japan), field emission scanning electron microscopy (FESEM) (LEO1530, Japan), high-resolution transmission electron microscopy (HRTEM) (JEOL-2010, Japan), and energy dispersive x-ray spectroscopy (EDS).

3. Results and discussion

Fig. 1(a) shows a representative SEM image of the obtained sample. The SEM image shows that the as-synthesized products consist of a large quantity of nanobelts with typical lengths in the range of several hundreds of micrometers. The width of the nanoblets is about 50-100 nm. XRD pattern taken from the pure and Mn-doped ZnS nanobelts were measured as shown in Fig. 1(b). The diffraction peaks can be indexed to the polycrystalline hexagonal wurtzite structure of ZnS from JCPDS (No.36-1450). From Fig. 1(b), as compared to the pure ZnO, it was also clear that the peak positions corresponding to the planes (002) and (101) were shifted to lower angle when Mn was incorporated, which is possibly related to the larger radius of Mn atoms than that of Zn atoms. Of all the diffraction patterns, two Au peaks from the Au catalyst on the substrate were detected and no other phases or impurities were observed, indicating that incorporation of Mn neither changed the wurtzite structure of ZnS nor resulted in the formation of Mn or Mn compound impurity phases.

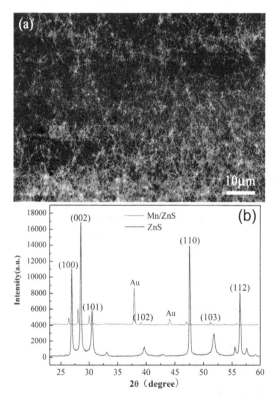

Fig. 1. (a) SEM image of the Mn-doped ZnS nanobelts grown on the silicon substrate. (b) XRD pattern taken from the pure and Mn-doped ZnS nanobelts.

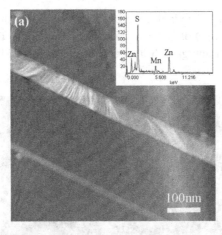

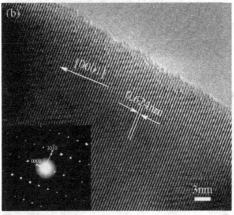

Fig. 2. (a) TEM image of a single Mn-doped ZnS nanobelt. The inset is the EDS of the nanobelt. (b) HRTEM image of the Mn-doped ZnS nanobelt. The inset is the corresponding SAED pattern.

The structure of the nanobelts was further characterized by HRTEM and EDS. Fig. 2(a) shows a typical TEM image of a single Mn-doped ZnS nanobelt. The inset in Fig. 2(a) shows a typical EDS pattern of the nanobelt, which consists of Zn, Mn and S. EDS measurement has been made on tens of nanobelts and the Mn content in most of them is identified to be about 5at% from the intensity ratios among the Zn, Mn and S peaks. Fig. 2(b) corresponds to the HRTEM image of the nanobelt, which indicates that the nanobelts are single crystalline and free of defects. The corresponding selected area electron diffraction (SAED) pattern confirms that the phase of the nanobelt is of the wurtzite ZnS structure. The lattice fringes with a separated spacing of 0.624 nm correspond to the (0001) plane of wurtzite ZnS. Combining the SAED patterns and HRTEM images, the growth direction of the nanobelts is confirmed to be along [0001].

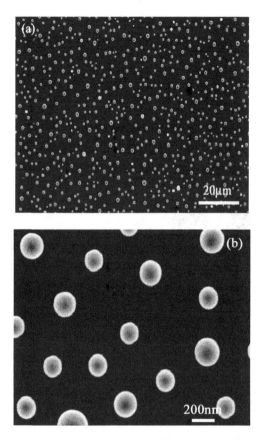

Fig. 3. Low (a) and High magnification (b) of SEM images of the Si substrate surface with Au catalysts after heat treatment

In order to investigate the growth mechanins of ZnS nanobelts, heat treatment process was carried out for the Si substrate with Au films. A furnace with a horizontal quartz tube was used to proceed the heat treatment. High-purity argon was kept flowing through the tube at a flow rate of 100 sccm to prevent surface oxidation of the Si substrate. The process was carried out at 1000°C for 2 hours and cooled to room temperature. Fig. 3 shows the SEM images of the substrate surface with Au catalysts after heat treatment. From the figure, we can see that the Au catalyst particles with the diameter of 100-200 nm are distributed homogeneously on the substrate. Au films melt to form catalyst particles due to the surface tension at high temperature.

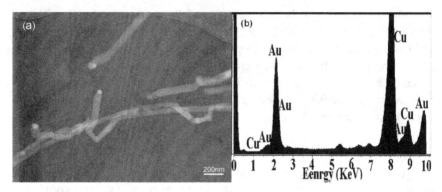

Fig. 4. TEM images of the nanobelts grown on the Si substrate with Au films (a) EDS pattern for the tip of the nanobelts (b).

Fig. 4(a) displays the TEM images of the nanobelts grown on the Si substrate with Au films. EDS pattern for the tip of the nanobelts shown in Fig. 4(b) demonstrate that Au particles were formed during the growth process. Cu comes from the Cu grid. Metal particles were found at the tip of the nanobelts in our TEM investigations, which indicate that the growth of the Mn doped ZnS nanobelts was governed by the commonly reported metal-seeded VLS mechanism [15,16].

Here we propose the growth mechanism of the Mn doped ZnS nanobelts. Although the melting point of pure gold and silicon is 1063 and 1412 °C, respectively, the eutectic temperature of the Au–Si system is known to be only 370 °C [17]. During the reaction process, the eutectic reaction between gold thin layer and the silicon substrate occurred firstly at 370°C and formed Au-Si eutectic alloys before zinc vapors beginning to release at approximately 400°C. At higher reaction temperature, zinc and sulfur vapors were generated. At the same time, $MnCl_2$ reacted with hydrogen via the reaction $MnCl_2+H_2 \rightarrow Mn+2HCl$. The Au-Si liquid alloy began to absorb the generated zinc and sulfur vapors and manganese atoms to reach equilibrium condition. When the liquid alloy became supersaturated, Mn doped ZnS nanobelts precipitated out and kept on growing from the supersaturated surface, possibly through the reactions $Mn(l)+Zn(l)+S(l) \rightarrow Mn/ZnS(s)$. Manganese ions incorporated into the lattice of ZnS by substituting parts of Zn atoms forming a single wurtzite structure.

4. Conclusions

The Mn doped ZnS nanobelts were successfully synthesized by H_2 assistant chemical evaporation deposition method at relatively low temperature. XRD and TEM analysis indicated that the doped nanobelts have wurtzite structure and grow along [0001] direction. The growth of the Mn doped ZnS nanobelts was governed by the metal-seeded VLS mechanism. The doped ZnS nanobelts are believed to be promising blocks for the construction of photoluminescent and electroluminescent nanodevices.

Acknowledgments

This work was supported by the Major Project of International Cooperation and Exchanges (2006DFB51000), NSFC (51172022), NSAF (10876001), the Research Fund of Co-construction Program from Beijing Municipal Commission of Education, the Fundamental Research Funds for the Central Universities.

References

1. E. Monroy, F. Omnes and F. Calle, Semicond. *Sci. Technol.* **18**, R33 (2003).
2. R. Menner, B. Dimmler and H.W. Schock, *J. Crystal Growth* **86**, 906 (1988).
3. R.N. Bhargava, D. Gallagher and D. Nurminkko, *Phys. Rev. Lett.* **72**, 416 (1994).
4. W. P ark, J.S. King, C.W. Neff, C. Liddell and C. Summers, *Phys. Status Solidi* **b229**, 949 (2002).
5. S. Velumani and J.A. Ascencio, *Appl. Phys. A: Mater. Sci. Process* **72**, 236 (2003).
6. E. Monroy, F. Omnes, F. Calle, *Semicond. Sci. Technol.* **18**, R33 (2003).
7. R.N. Bhargava, D. Gallagher, X. Hong, D. Nurminkko, *Phys. Rev. Lett.* **72**, 416 (1994).
8. W. Chen, R. Sammynaiken, Y. Huang, J.O. Malm, R. Wallenberg, J.O. Bovin, V. Zwiller and N.A. Kotov, *J. Appl. Phys.* **89**, 1120 (2001).
9. A.D. Dinsmore, D.S. Hsu, S.B. Qadri, J.O. Cross, T.A. Kennedy, H.F. Gray and B.R. Ratna, *J. Appl. Phys.* **88**, 4985 (2000).
10. C.J. Barrelet, Y. Wu, D.C. Bell and C.M. Lieber, *J. Am.Chem. Soc.* **125**, 11498 (2003).

11. Y.W. Wang, L.D. Zhang, C.H. Liang, G.Z. Wang and X.S. Peng, *Chem. Phys. Lett.* **357**, 314 (2002).

12. Y. Jiang, X.M. Meng, J. Liu, Z.R. Hong and S.T. Lee, *Adv. Mater.* **15**, 1195 (2003).

13. X.H. Zhang, Y. Zhang, Y.P. Song, Z. Wang and D.P. Yu, *Physica E* **28**, 1 (2005).

14. J.P. Ge, J. Wang and H.X. Zhang, *Adv. Funct. Mater.* **15**, 303~308 (2005).

15. M. H. Huang, S. Mao, H. Feick, H. Yan, Y. Wu, H. Kind, E. Weber, R. Russo and P. Yang, *Science* **292**, 1897 (2001).

16. A. M. Morale and C. M. Lieber, *Science* **279**, 208 (1998).

17. T. Adachi, *Surface Science* **506**, 305 (2002).

SHAPE-DEPENDENT SURFACE PLASMON RESONANCE OF AG NANOCRYSTALLINES IN OPAA TEMPLATE

SHANSHAN HAN

School of Material Science and Engineering, Tongji University, Shanghai 200092, P. R. China

XIU-CHUN YANG

School of Material Science and Engineering, Tongji University, Shanghai 200092, P. R. China

Transparent Ag/OPAA composite films were prepared by filling silver nanocrystallines (NCs) into an ordered porous anodic alumina (OPAA) templates using a direct current (DC) deposition technique. The broad and significant absorption peak in visible light region for each transparent Ag/OPAA composite could be attributed to surface plasmon resonance (SPR) of Ag NCs. The SPR peak could be separated into three peaks by Lorentzian fits, which corresponded to transverse quadrupole resonance, transverse dipole resonance and longitudinal resonance, respectively. The influences of size, shape and volume fraction of Ag NCs on the maxima and intensities of SPR peaks were discussed.

1. Introduction

When metal nanocrystallines (NCs) are irradiated by light, the oscillating electric field causes the free conduction electrons to oscillate coherently, which is denoted as surface plasmon resonance (SPR) [1]. Since the establishment of Mie's theory in 1908, a mass of studies [2-5] have been reported to understand the nature and the influence factors of SPR due to its important applications in areas such as surface enhanced Raman scattering (SERS) [6-8], nonlinear optics [9-11], and photonic crystals [12].

Recently, Zong et al. [13] reported the SPR properties of Ag nanowire arrays filled in OPAA template by alternating current (AC) electrodeposition. Traditionally, in order to fill completely OPAA by electrodeposition, the porous anodic alumina needs to be detached from the aluminum substrate and the barrier layer must be removed from the porous alumina by a chemical etching process, a metallic electrode is afterwards sputtered on one side of the porous anodic alumina membrane. For optical research, the metallic electrode must be removed. The complexity of the process severely limited its optical application. Direct current (DC) electrodeposition was seldom used to fabricate metal/OPAA composites for optical research because it was difficult for the electrons to tunnel through the barrier layer [14]. Therefore a new procedure was developed to electrochemically fill OPAA where porous alumina remained on the aluminum substrate and the barrier layer was largely thinned by using a step-by-step voltage decrement process [15] or a twice constant current anodization process [16]. The thinning leaded to a considerable decrease in the potential barrier for the electrons to tunnel through the barrier layer, when the metal was deposited at the pore tips. Using DC electrodeposition, we have prepared 30-μm-long Cu and Ag nanowire arrays [17, 18]. However, due to the great length of Cu and Ag nanowires, the composites are optically opaque, which makes the optical property studies of the composite difficult.

Herein, we successfully synthesized transparent Ag nanocrystallines/OPAA composites film by this modified DC electrodeposition method and investigated the influences of size, shape and volume fraction of Ag nanocrystallines on its SPR maxima and intensities.

2. Experimental

2.1. *Preparation of transparent Ag nanowires arrays*

A highly ordered and large-area OPAA template was fabricated by a two-step anodization process plus a step-by-step voltage decrement method as described previously [17, 18].

Electrodeposition was performed on LK98II electrochemical system (Lanlike, China). In the electrodeposition cell, the OPAA template with Al substrate, Pt plate and saturated calomel electrode (SCE) were used as the working electrode, the counter electrode and the reference electrode, respectively. Constant voltage (-6.5 V) DC electrochemical deposition was employed in a mixing electrolyte of 0.01 mol/L AgNO$_3$ and 0.1 mol/L H$_3$BO$_3$, here H$_3$BO$_3$ was used as buffer reagent. Samples S1, S2 and S3 were electrochemically deposited for 10 s, 40 s and 80 s, respectively. After deposition, the as-prepared samples were rinsed with deionized water, and then the Al substrate was removed by CuCl$_2$ solution.

40

2.2. *Characterization*

Optical photographs of the OPAA template and the Ag/OPAA film were taken by Sony camera. Hitachi 3310 UV-Vis spectrophotometer was used to measure optical absorption spectra, and an unpolarized light beam at normal incidence to the film plane was used for the measurements. Quanta 200 FEG scanning electron microscope (FESEM) with an energy-dispersive x-ray spectroscope (EDS) was used to characterize the morphology and elemental composition. H-800 transmission electron microscope (TEM) was used to analyze the morphology and microstructure of these samples. TEM samples were prepared by immersing a small piece of Ag/OPAA composite in 2 mol/L NaOH solution for about 5 h (60 ℃) in order to dissolve the OPAA template. Afterwards, Ag NCs were separated out of the solution by centrifugal effects. Finally, the deposit was ultrasonically dispersed in 3–5 mL ethanol, and a drop of the suspended solution was placed on a Cu grid with carbon membrane for TEM observation.

3. Results and discussion

Figure 1 presents the photos of OPAA templates before and after electrodepositing silver. The background is a piece of white paper printed with the logo of Tongji University.

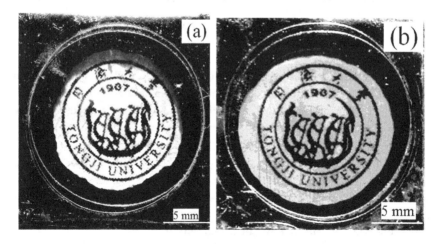

Figure 1. Camera photos of OPAA template (a) and sample S1 (b) with the logo under these samples

As shown in Figure 1 (a), the opaque border is Al matrix, which can be used as the framework to support the brittle OPAA template. The inner region is the

OPAA template. The logo under the OPAA template can be seen clearly, indicating that the OPAA template is virtually transparent. Figure 1(b) shows that the OPAA template becomes orange-red after depositing Ag NCs, and the logo under the composite can be seen clearly, indicating that the filled composite is still transparent. Extending the electrochemical deposition time to 80 s, the composite becomes greenish-brown and is still transparent.

Figure 2 gives FESEM photographs and EDS spectra of sample S1.

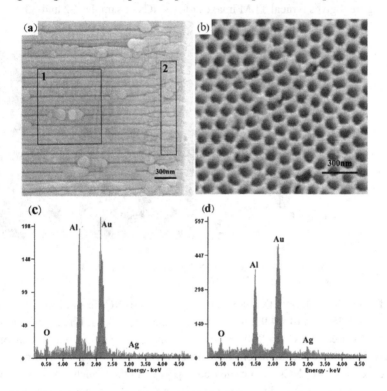

Figure 2. FESEM photographs and EDS spectra of S1: the cross-section image (a), the top-view (b), EDS spectra for region 1 (c) and region 2 (d)

Figure 2(a) indicates that the pore channels can be divided into an ordered pore layer as shown in region 1 and a branched pore layer as shown in region 2. In the ordered pore layer, the pore channels are straight and parallel to each other. The straight pores branch out at the formation front because the pore diameter is proportional to the anodizing potential and the pore density is inversely proportional to the square of the anodizing potential [17, 19]. The thickness of the branched pore layer is about 500nm. Figure 2(b) is the top-view of the OPAA template, which shows that the honeycomb-like template is highly ordered with circular holes and hexagonal structure cell. The ordered pore diameters range from 88 nm to 98 nm. No silver can be found in region 1 as shown in Figure 2(c). Figure 2(d) indicates the existence of Ag atoms in region 2. Al is from the OPAA template, and Au is from the Au film deposited on the observed surface.

Figure 3 gives typical TEM images of Ag NCs of samples S2 and S3.

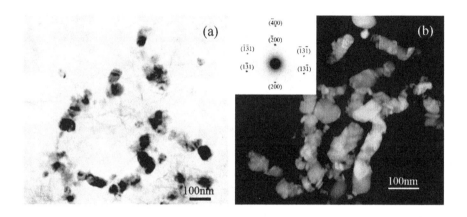

Figure 3. TEM images of Ag NCs in samples S2 (a) and S3 (b).

Figure 3(a) indicates that the diameters of the Ag NCs in samples S2 are 45-55 nm with lengths of 60-90nm. With increasing deposition time, the length of the Ag NCs can be up to 200 nm without changing the diameter as shown in Figure 3(b). Since the diameter of the ordered pore channels is 88-98 nm, it can be deduced that Ag NCs should only be deposited in the branched pore channels, which coincides with EDS results in Figure 2. The selected area electron diffraction (SAED) of sample S3 inserted in Figure 3(b) indicates that Ag NC is monocrystalline with face center cubic structure.

Figure 4(a) gives optical absorption spectra of OPAA template, samples S1, S2 and S3. Figure 4(b) gives Lorentzian fits for the experimental spectra of samples S1 and S3.

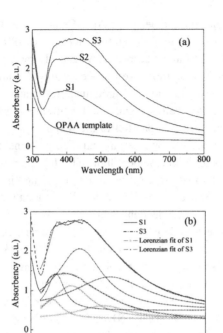

Figure 4. Optical absorption spectra of OPAA template and samples S1, S2, S3 (a) and Lorentzian fits of samples S1 and S3 (b)

For the empty OPAA template, its optical absorption is very weak at the wavelength longer than 350 nm, indicating that it can be an excellent matrix for fabrication of optical devices. For samples S1, S2 and S3, noticeable and broad absorption peaks have been observed, which can be denoted as the SPR absorption of Ag NCs [1-2].

As well known, SPR absorption is affected by size, shape and volume fraction of the metal NCs [1-5]. When the diameter of spherical metal NCs are much smaller than the wavelength of the exciting light ($\lambda \geqslant 20R$), only dipole resonant mode contributes to the absorption spectrum. When the spherical NCs become larger and comparable to the wavelength of the exciting radiation, inhomogeneous polarization of metal NCs emerges. Some higher-order multipole resonant modes, especially quadrupole resonance, become important to the absorption spectrum. Furthermore, when the spherical NCs change to be nanorods, the oscillation of the free electrons perpendicular to the long axis of the nanorods must be taken into account [20].

In our experiment, the diameters of Ag NCs are larger than 45nm and the aspect ratios range from 1 to 5. Therefore, the broad SPR spectrum of each sample could be divided into three peaks by using Lorentzian fits as shown in Figure 4. (b): transverse quadrupole resonance around 360 nm, transverse dipole resonance around 420 nm and longitudinal resonance around 520 nm, respectively [4, 13]. As well known, the longitudinal resonance cannot be excited when the light beam is parallel to the major axis of the nanorods. The existence of longitudinal resonance around 520 nm indicates that the Ag nanorods are not parallel to the incident light. This has been demonstrated by the above-metioned results that Ag NCs were only deposited in the branched pore channels, which are not parallel to the incident light though the pore channels in region 1 are parallel to the incidendent light.

With increasing electrodeposition time, Ag volume fraction in the OPAA template increases, which induces an enhanced SPR peak. This is in good agreement with Mie's theory, which predicts a proportional relationship between metal volume fraction and intensity of SPR absorption.

For the transverse dipole resonance, the fitted peak in sample S3 has a little red shift compared to that in sample S1. This is consistent with Link's reports that the maximum of the transverse dipole resonance absorption shifts to longer wavelength with increasing NCs' size, especially for large NCs [2, 3].

For the transverse quadrupole resonance peaks, maxima shift to longer wavelength with increasing deposition time. It could also be explained as the size dependence effort [2, 3].

The aspect ratio of metal NCs plays an important role to affect the longitudinal resonance [20, 21]. For sample S1, the peak of longitudinal resonance is very weak, because most Ag NCs in sample S1 are nearly spherical due to the shorter deposition time. For sample S3, with prolonging deposition time, the volume fraction and aspect ratio of Ag nanorods increases as demonstrated by TEM photographs, accordingly, the longitudinal resonace enhances and red-shifts, which is consistent with to Gans theory that the maximum of longitudinal mode shifts to longer wavelength with increasing aspect ratio of the nanorods [20, 21].

4. Conclusions

Optically transparent Ag NCs/OPAA composites are successfully fabricated by constant voltage DC electrodeposition. Ag NCs/OPAA composite shows a significant SPR absorption, which can be divided into transverse quadrupole resonance, transverse dipole resonance and longitudinal resonance. Most Ag

NCs in sample S1 are sphere, and the volume fraction of Ag nanorods increases with increasing deposition time, accordingly, the relative intensity ratio of longitudinal to transverse dipole resonance becomes larger. All of the resonance modes have red-shifts and their intensities enhance with increasing deposition time.

Acknowledgements

This work was supported by National Natural Science Foundation of China (No.50672069), Key Project for Basic Research of Shanghai (08JC419000) and the Nanotechnology Special Foundation of Shanghai (No.11nm0500700).

References

1. U Kreibig, *M Vollmer*, **Springer:** Berlin, (1995).
2. S Link, E M A l –Sayed, *J. Phys. Chem. B* **103**, 4212 (1999).
3. S Link, M A El –Sayed, *J. Phys. Chem B* **103**, 8410 (1999).
4. K L Kelly, E Coronado, L Zhao, et al. *J. Phys. Chem. B* **107**, 668 (2003).
5. Thomas S, Nair S K, Jamal E M A, et al. *Nanotechnology* **19**, 075710 (2008).
6. H B Chu, J Y Wang, L Ding, et al. *J. Am Chem. Soc.* **131**, 14310 (2009).
7. N Ji, W D Ruan, C X Wang, et al. *Langmuir,* **25**, 11869 (2009).
8. B Ren, X F Lin, Z L Yang, et al. *J. Am. Chem., Soc.* **125**, 9598 (2003).
9. Jiang Y, Wang H Y, Xie L P, et al. *J. Phys. Chem., C* **114**, 2913 (2010).
10. Yang X C, Dong Z W, Liu H X, et al. *Chem., Phys. Lett.,* **475**, 256 (2009).
11. J L Gu, J L Shi, G J You, et al. *Adv. Mater.,* **17**, 557 (2005).
12. K Kaneko, K Yamamoto, S Kawata, et al. *Opt. Lett.* **33**, 1999 (2008).
13. R L Zong, J Zhou, Q Li, et al. *J. Phys. Chem. B* **108**, 16713 (2004).
14. Y T Pang, G W Meng, L D Zhang, et al. *Nanotechnology* **14**, 20 (2003).
15. J Choi, G Sauer, K Nielsch, et al. *Chem. Mater.* **15**, 776 (2003).
16. K Nielsch, F Müller, A P Li, et al. *Adv. Mater.* **12**, 582 (2000).
17. X C Yang, X Zou, Y Liu, et al. *Mater. Lett.,* **64**, 1451 (2010).
18. X Zou, X N Li, X C Yang, et al. *Journal of Functional Materials (Chinese)* **41**, 321 (2010).
19. A P Li, F Müller, A Birner, et al. *J. Appl. Phys.* **84**, 6023 (1998).
20. R Gans. Ann *Phys,* **47**, 270 (1915).
21. X C Yang, H X Liu, L L Li, et al. *Journal of Functional Materials (Chinese)* **41**, 341 (2010).

ELECTRICAL PROPERTIES AND ELECTRIC-INDUCED NANODAMAGE OF SINGLE CRYSTALLINE BEAD-SHAPED ZNO NANOROD

HUIFENG LI, PEIFENG LI, YUNHUA HUANG[§]

[1]Department of Materials Physics, University of Science and Technology Beijing, 30 Xueyuan Road, Beijing 100083 China
[2]State Key Laboratory for Advanced Metals and Materials, University of Science and Technology Beijing, 30 Xueyuan Road, Beijing 100083 China

Bead-shaped ZnO nanorods (NRs) are synthesized through the direct evaporation of metal zinc and graphite powders in Ar and O_2 with Au catalyst. The electrical properties and electric-induced nanodamage of single crystalline bead-shaped ZnO NRs have been studied by nano-manipulation and measurement systems in scanning electron microscopy (SEM). Electrical properties reveal that the different contact between the electrodes and the NRs will impact the *I-V* characteristics of ZnO NRs. Current-induced breakdown caused by Joule heating has been achieved by applying suitably high voltages. Our study will be helpful for predicting the reliability of the nanoelectromechanical systems (NEMS) devices made of ZnO NRs

1. Introduction

As a wide and direct band gap (3.37 eV) semiconductor with a large exciton binding energy (60 meV), ZnO has been widely used in piezoelectric transducers, piezoelectric gated diodes [1-2], field-effect transistors [3-4], strain sensors [5-6], and nanogenerators [7-9]. In many types of ZnO nanostructures, such as nanowires, nanorods, nanocombs, nanobelts, nanorings, nanowires represent the key building blocks for bottom-up assembly of complex functional architectures for future electronic and optoelectronic nanodevices. The electrical properties of individual ZnO nanowire have been studied mainly in the context of field effect transistors [3, 10], and the concentration and mobility of carriers in nanowires have been estimated. Asymmetric behaviors [11] and bending features [12] caused by contact between ZnO nanowires and metal electrodes have been characterized using *I–V* curves. And effective contact between nanowires and electrodes is important for achieving nanoscale electronic devices. Furthermore, the prototype NEMS devices based on ZnO nanowires

[§] E-mail: huangyh@mater.ustb.edu.cn

have been developed, and the investigation about electrical and mechanical reliability of NEMS devices becomes more and more important and urgent because all of the NEMS devices have to face this problem. All practical applications based on NEMS devices require the knowledge of the safe service conditions in electrical environments [13]. Currently, our group reported the electrical and mechanical coupling nanodamage in single ZnO nanobelt through a conductive atomic force microscope [13], and electrical breakdown of ZnO nanowires when the applied electric field reached the break point ~10^6 V/m by nano-manipulation systems [14]. Similarly, Bando's group reported the direct observation of thermal decomposition of boron nitride nanotubes (BNNTs) under current flow and studied the dependence of thermal decomposition temperature on local electrical field [15]. Beaded ZnO nanorods are unique morphology in the family of ZnO nanostructures, and there were few reports on the electrical properties and breakdown in single beaded ZnO nanorod. The significance of the investigation of electrical nanodamage is that it can provide the designers of NEMS devices with useful material property parameters, which can be used to evaluate the electrical and mechanical reliability of nanodevices and predict the safe service conditions of devices.

In this paper, besides the fabrication and characterization of bead-shaped ZnO NRs, we focus on nano-manipulation and measurement systems based technique for investigating the electrical failure in tungsten-ZnO NR-tungsten (W-ZNR-W) structure. This technique is convenient and can also be used to investigate the electrical nanodamage of other nanomaterials. This study will be helpful for predicting the reliability of the NEMS devices made of ZnO NRs.

2. Experimental

ZnO bead-shaped NRs were synthesized successfully using pure Zn (99.99%) and graphite powders with molar ratio of 8:1 by CVD. The flow rate of argon and oxygen was 50 and 5 standard cubic centimeters per minute (sccm) separately. The quartz tube was heated up to 550 °C, and retain reaction for 30 minutes. Finally, a thin layer was deposited on the silicon wafer.

The morphologies and structures of the synthesized product were characterized using SEM (JEOL-6490, Japan), high-resolution transmission electron microscopy (HRTEM) (JEOL-2010, Japan), and selected area electron diffraction (SAED). The I-V characteristics of the individual bead-shaped ZnO NR were measured by the nano-manipulation and measurement systems (Zyvex S100 plus Keithley 4200-SCS, American) in SEM.

3. Results and Discussion

3.1. *Morphology and Structures*

Figure 1a shows a typical SEM image of the as-synthesized beaded nanorods with an average diameter of 100-500 nm and a length of several tens of microns. Figure 1b shows that the individual beaded ZnO has a smooth surface and indicates the beaded ZnO NR cyclical changes, in which many pearls cluster as a line and each pearl is oval shape with a diameter of 200-400 nm and length of 300-600 nm. Further structural characterization of individual beaded ZnO NR is performed by TEM, as shown in Figure 1c, from which we can see that the ZnO NR has the characteristic beaded morphology as same as that observed by SEM. Figure 1d is a high-magnification TEM image, and the inset shows the corresponding SAED pattern taken from the nanorod. By combining the HRTEM images with the SAED pattern, the growth direction of the fraction can be determined along $[10\bar{1}0]$ and $[2\bar{1}\bar{1}0]$.

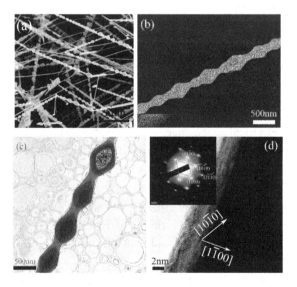

Figure 1. (a) Low-magnified and (b) high-magnified SEM image of beaded ZnO NRs; (c) Low-magnified TEM and (d) HRTEM image of individual beaded ZnO NR, inset: SEAD image of beaded ZnO NR.

3.2. *Electrical Properties*

We studied the *I-V* characteristic of beaded shape ZnO NRs with the W tip contacting with the ends of the NR, as shown in figure 2a. It can be seen from the image that the single bead-shaped ZnO NR with diameter of about 100-1000 nm and length about 50 μm. Figure 2b shows the *I-V* curve of beaded ZnO NR at voltage ranges from -5 to 5 V. The results indicate that the *I-V* curve is typical Schottky contact characterization with positive turn-on voltage about 2.5 V. As we all know, the metal-semiconductor (MS) contact can either be an Ohmic contact or a Schottky barrier depending on the Fermi surface alignment and the nature of the interface between the metal and the semiconducting NW [16]. The different contact between the electrodes and the NR will impact the *I-V* characteristics of ZnO NW. Then, to obtain a good Ohmic contact, Pt as electrodes was fabricated using a FIB microscope. The SEM image is shown in figure 2c, in which the single ZnO NR connects two Pt electrodes and the W tip directly connects Pt electrodes. The insert shows the end of beaded ZnO NR. We can firmly fixed the ZnO NR by FIB and reduce the contact resistance between the W tip and ZnO NR. Figure 2d shows the *I-V* curve with an Ohmic behavior that the current increases linearly with the bias scanning from -1 to 1 V. From the slope, diameter and length of the NR are measured from the SEM image, the conductivity of the beaded ZnO NR is measured to be 0.0667 $\Omega^{-1} \cdot cm^{-1}$.

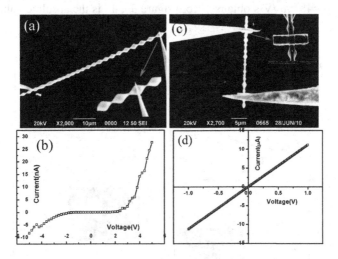

Figure 2. Typical SEM image of the measured beaded ZnO NR (a) directly contact with W tip and (b) Pt electrode contact; (c) and (d) is corresponding *I-V* curves of the beaded ZnO NR.

3.3. *Electric-induced Nanodamage*

The electric-induced nanodamage of single ZnO NR was measured with a nano-manipulation system using the W tip. Figure 3a and 3b demonstrate the SEM images of the two terminal electrical measurements on individual NR. Figure 3a shows W-ZNR-W applied a sweep voltage from 0 to 20 V. The experimental process of the W-ZNR-W system and the corresponding *I-V* characteristics are given in figure 3a. The results reveal a rectifying behavior because the work function of W is 4.55 eV and the electron affinity is 4.3 eV for ZnO, and the rectifying behavior originates from the Schottky contacts at the metal/ZnO interface. As the applied voltage range increased from 0 to 30 V, the NR fractured at the interface between the beads. The insert of figure 3b shows the final state. The *I-V* curve declined after the destruction happened, as illustrated in figure 3b. When failure took place, the current also dropped. Here, the turning point of the *I-V* curve in the positive voltage region is not the discontinuity point because thermionic emission is defined as the break-point and the corresponding voltage is the break-voltage. For this sample, the break-voltage obtained was 28 V and the electron mobility was about 48 cm^2/V·s according to the *I-V* curve displayed in figure 3b and the following relationship [14]:

$$\mu_0 \, |E| \approx u \tag{1}$$

where u is the sound velocity, 2.7×10^5 cm/s for ZnO [17]; μ_0 is the carrier mobility, $\mu_0 \approx 48$ cm^2/V·s obtained from figure 3b, |E| is the absolute value of the applied electric field, and $E = 5.6 \times 10^5$ V/m for this experiment.

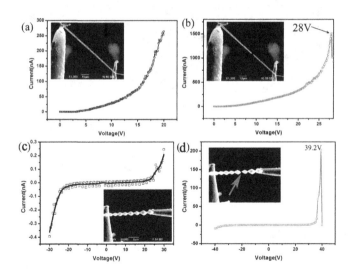

Figure 3. (a), (c) *I-V* curve of the single beaded ZnO NR before its destruction, SEM image of measurement; (b), (d) *I-V* curve recorded during the failure process. SEM images of the broken beaded ZnO NR.

The same experiments were performed on another beaded shape ZnO NR with different diameter, as shown in figure 3c and 3d. Figure 3c shows the *I-V* characteristics of the W-ZNR-W system applied a sweep voltage from -30 to 30 V. The experimental SEM image is given in the insert of figure 3c. It can be seen from the image that the beaded ZnO NR has a length about 15 μm, and contacts the two ends of W tip. The results indicate that the *I-V* curve was typical Schottky contact characterization with the positive turn-on voltage about 23 V. When the voltage range increased to ±40 V, the bead-shaped ZnO NR was broken down at the break-point of 39.2 V, as shown in Fig. 5b. The electron mobility was about 12.4 cm^2/V·s at the break-point using the previous method. The failure process was confirmed by several times and the current heating was considered to be the major factor for failure in nanostructures [15].

Figure 3a and 3c show the *I-V* curves of the beaded ZnO NR, which exhibit nonlinear and asymmetric behavior. Figure 4 shows the plot of the *I-V* curve in log scale in the range from about 15 to 20 V (corresponding Fig. 3a) and 24 to 30 V (corresponding Fig. 3c), respectively, which shows linear behavior. According to the detailed discussion by Zhang et al. [18], this linear behavior indicates that the ZnO NR makes two Schottky Barrier (SB) contacts with the two W tips, a back-to-back SB structure is formed, and the current is dominated by the tunneling current of the reverse biased SB, which follows the expression,

$$\ln I = \ln(SJs) + V(q/kT - 1/E_0) \qquad (2)$$

where, S is the contact area of SB, Js is the reverse saturation current density, k is the Boltzmann constant, T is the absolute temperature, q is the magnitude of electronic charge, V is the applied reverse bias, and E_0 is a parameter that depends on the carrier density. This logarithmic plot of the current *I* as a function of the bias V gives approximately a straight line of slope $q/(kT) - 1/E_0$. Electron concentration n can then be obtained via E_0 and the electron mobility can be calculated using the relation $\mu = 1/(nq\rho)$, with ρ being the resistivity of the NR. From the slop of the fitted line in Fig. 4a and 4b, E_0 is obtained as 25.7 and 26.1 mV, respectively. According to the relationship between E_0 and donor concentration N_D, N_D is obtained as 0.8×10^{16} cm^{-3} and 3.4×10^{16} cm^{-3}, respectively [13].

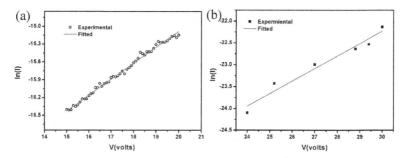

Figure 4. Experimental and fitted ln(I) vs V plot for ZnO using the *I-V* curve shown in Fig. 3a and 3c, respectively.

4. Conclusions

In summary, we synthesized bead-shaped ZnO NRs via a thermal evaporation method, and the structure characterization revealed that the individual beaded ZnO NR is made up of many pearls in line with smooth oval surface and cyclical changes. The *I-V* feature demonstrated that the MSM contact can either be Ohmic contact or a Schottky barrier depending on the Fermi surface alignment through changing the contact between ZnO NR and metal W tip. Moreover, the Joule heat will lead to fracture the ZnO NRs at the high voltage, and the break point is different as the difference of diameter of nanorods. According to the thermionic emission theory, the current is dominated by majority carriers, and the hot electrons acquire more energy when the applied electric field exceeds the break point. Furthermore, these results will support the applications and safe service of ZnO NRs as optoelectronics and NEMs devices.

Acknowledgments

This work was supported by the Major Project of International Cooperation and Exchanges (2006DFB51000), NSFC (51172022), NSAF (10876001), the Research Fund of Co-construction Program from Beijing Municipal Commission of Education, the Fundamental Research Funds for the Central Universities.

References

1. J. H. He, C. L. Hsin, J. Liu, L. J. Chen and Z. L. Wang, *Adv. Mater.* **19**, 781 (2007).
2. Y. Yang, J. J. Qi, Q. L. Liao, H. F. Li, Y. S. Wang, L. D. Tang and Y. Zhang, *Nanotechnology*, **20**, 125201 (2009).

3. X. D. Wang, J. Zhou, J. H. Song, J. Liu, N. S. Xu and Z. L. Wang, *Nano Lett.* **6**, 2768 (2006).

4. J. Goldberger, D. J. Sirbuly, M. Law and P. D. Yang, *J. Phys. Chem. B* **109**, 9 (2005).

5. J. Zhou, Y. D. Gu, P. Fei, W. J. Mai, Y. F. Gao, R. S. Yang, G. Bao and Z.L. Wang, *Nano Lett.* **8**, 3035 (2008).

6. Y. Yang, J. J. Qi, Y. Zhang, Q. L. Liao, L. D. Tang and Z. Qin, *Appl. Phys. Lett.* **92**, 183117 (2008).

7. Z. L. Wang, J. H. Song, *Science*, **312**, 242 (2006).

8. G. Zhu, R. S. Yang, S. H. Wang and Z. L. Wang, *Nano Lett.* **10**, 3151 (2010).

9. S. Xu, Y. Qin, C. Xu, Y. G. Wei, R. S. Yang and Z. L. Wang, *Nat. nanotech.* **5**, 366 (2010).

10. H. F. Li, Y. H. Huang, X. J. Xing, J. Su and Y. Zhang, *Materials Science Forum*, 654-656 (2010).

11. Y. F. Hu, Y. L. Chang, P. Fei, R. L. Snyder and Z. L Wang, *ACS Nano* **4**, 1234 (2010).

12. K. H. Liu, P. Gao, Z. Xu, X. D. Bai and E. G. Wang, *Appl. Phys. Lett.* **92**, 213105 (2008).

13. Y. Yang, J. J. Qi, Y. S. Gu, W. Guo and Y. Zhang, *Appl. Phys. Lett.* **96**, 123103 (2010).

14. Q. Zhang, J. J. Qi, Y. Yang, Y. H. Huang, X. Li and Y. Zhang, *Appl. Phys. Lett.* **96**, 253112 (2010).

15. Z. Xu, D. Golberg and Y. Bando, *Nano Lett.* **9**, 2251 (2009).

16. K. Subannajui, D. S. Kim and M. Zacharias, *J. Appl. Phys.* **104**, 014308 (2010).

17. H. Kawamura, H. Yamada, M. Takeuchi, Y. Yoshino, T. Makino and S. Arai, *Vacuum* **74**, 567 (2004).

18. Z. Y. Zhang, C. H. Jin, X. L. Liang, Q. Chen and L. M. Peng, *Appl. Phys. Lett.* **88**, 073102 (2006).

OPTICAL PROPERTIES AND PHOTOCATALYTIC ACTIVITY OF MN-DOPED ZNO NANORODS

JING ZHAO, ZHIMING BAI, XIAOQIN YAN[*]

Department of Materials Physics and Chemistry, School of Materials Science and Engineering, University of Science and Technology Beijing, Beijing, 100083, China

LI WANG[*]

School of Civil & Environmental Engineering, University of Science and Technology Beijing, Beijing, 100083, China

We have successfully fabricated pure ZnO nanorods and Mn-doped ZnO nanorods with different concentrations, which are 1%, 2.5%, and 5%. XRD and Raman scattering measurements were performed to investigate the structures of Mn-doepd ZnO nanostructres. It was found that the Mn atoms tended to substitute for O atoms till reach a saturation point. Absorption spectra were taken to derive the width of optical band gap of Mn-doped ZnO nanorods. The doping of Mn element can make the band gap firstly increase and then decrease. The photocatalytic activity of samples was studied by comparing the degradation rate of rhodamine B (RB) under UV-light irradiation.

1. Introduction

Zinc oxide is a wide band gap semiconductor (3.37 eV) with large exciton binding energy of 60 meV at room temperature [1]. In recent years, ZnO nanostructures have attracted much attention for their potential applications in photocatalytic area. It is claimed by early reports that doping of transitional metal could further enhance the photocatalytic activity of ZnO by inhibiting the recombination of exciton and tuning the width of band gap [2, 3]. But recently, it makes the statement more debatable that many reports appear saying the transitional metal doping could sometimes reduce the photocatalytic acticity of ZnO photocatalysis.

For instance, K. C. Barick et al reported the doping of Mn, Co, and Ni could all reduce the photocatalytic efficiency of ZnO under ultra-violet (UV) light, because the transition metal ions substituted in ZnO lattice could inhibit

---

[*] Corresponding authors, e-mail:xqyan@mater.ustb.edu.cn, wangli@ces.ustb.edu.cn

the movement of electrons and holes [4]. Besides, Xiaoqing Qiu et al reported Co-doping could enhance the photocatalytic activity of ZnO under visible light, but the Co-doping could also cause a reduction in photodegradation efficiency under UV light for the reason that Co dopants made an increase in the number of trapping or recombination centers for excitons [5].

The cause of the conflicting conclusions is that the change of photocatalytic activity is determined by a good many factors including positive factors and negative factors. So, it becomes very necessary and important to study on every factor which can affect the photocatalytic activity.

In this paper, we successfully fabricated Mn-doped ZnO nanorods with different concentrations, which are around 0%, 2%, and 5%. XRD and Raman scattering measurements were performed to investigate the structures of Mn-doepd ZnO nanorods. Absorption spectra were taken to derive the change of optical band gap width of ZnO nanorods after Mn-doping. The photocatalytic activity of ZnO samples under ultra-violet light was studied by analyzing the degradation of rhodamine B (RB).

2. Experimental Section

2.1. Preparation of Mn-doped ZnO nanorods

The Mn-doped ZnO nanorods were prepared through a chemical route. Zinc acetate $[Zn(Ac)_2 \cdot 2H_2O]$ (5mmol) and required amount of nickel acetate $[Ni(Ac)_2 \cdot 4H_2O]$, which are 0, 0.1, and 0.25 mmol, were dissolved into 175ml absolute ethanol, then 40mmol NaOH and 40.0ml PEG-400 were added into the above solution. The reaction was kept at 140 ℃ for 16h in a Teflon-lined stainless autoclave tank. The obtained brown precipitates after reaction were washed, then filtered, and finally dried in air at 60 ℃ for 4h.

2.2. Characterization

The morphologies and component of as-prepared ZnO products were characterized by field emission scanning electron microscopy (FE-SEM) (Zeiss, SUPRA-55) equipped with an energy-dispersive x-ray spectrometer (EDX). The crystallinity of the samples were determined by x-ray diffraction (XRD) technique using D8 discover with GADDS and Raman spectroscope meter (Jobin-Yvon, HR800) with the 514 nm line of an Ar^+ laser. The absorption spectra were recorded with a Hitachi spectrophotometer (U-3010).

2.3. Measurement of photocatalytic activity

Rhodamine B (RB) was a widely used organic matter to evaluate the photocatalytic activity of ZnO [6-9]. The concentration of RB solution used in our work is $1 \times 10^{-5}M$ (100mL), which is contained by a 200mL quartz beaker.

56

2mg Mn-doped ZnO samples were dispersed into the beaker under magnetic stirring, and kept in dark for 30 minutes to reach the adsorption equilibrium. Then the dispersion was irradiated by a 300W UV-light mercury lamp (Osram) with the central wavelength of 365nm, and the surface of the dispersion was maintained a distance of 15cm from the light source to avoid heat effect.

3. Results and Discussions

3.1. *Morphology and component*

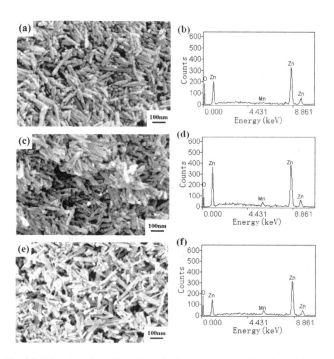

Figure 1. The SEM image and EDX spectra of Mn-doped ZnO nanorods; (a) (b) 1% Mn; (c) (d) 2.5% Mn; (e) (f) 5% Mn.

Figure 1 shows the morphologies of Mn-doped ZnO nanorods. Although the Mn contents are different, the sizes of ZnO nanorods are nearly the same, which are 15-20 nm in diameter and 80-200 nm in length. The Mn contents in ZnO nanorods are 0%, 2.5%, and 5%, which are obtained by EDX analyzing (in insets of Figure1), and the test were repeated at three different spots for each sample to make sure the distribution of Mn was homogeneous.

3.2. Structure characterization

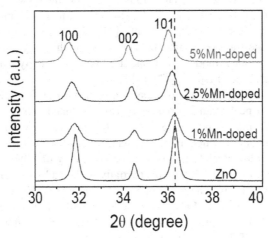

Figure 2. XRD patterns of pure ZnO and Mn-doped ZnO.

Figure 2 shows the enlarged XRD patterns around (100), (002), and (101) peaks of ZnO samples. No extra peak was found in X-ray diffraction (XRD) patterns besides the peaks of ZnO wurtzite structure, suggesting no secondary phase in Mn-doped ZnO. The XRD peaks of Mn-doped ZnO nanorods are shifted to lower diffraction angles compared with those of pure ZnO. Because the ionic radius of Mn^{2+} (0.90Å) is larger than that of Zn^{2+} (0.76Å), the substitution would certainly result in the expansion of the lattice [10].

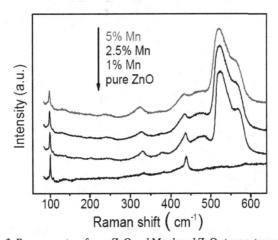

Figure 3. Raman spectra of pure ZnO and Mn-doped ZnO at room temperature.

The Raman spectra further illustrate the structure of Mn-doped ZnO nanorods, as shown in Figure 3. The two peaks at 100 and 438cm⁻¹ are attributed to the first-order low-frequency E_2 mode and high-frequency E_2 mode, which are the characteristic modes of wurtzite ZnO [11]. The weak peak at 380cm⁻¹ is indicated to be the transverse optical (TO) mode with A_1 symmetry, which is also a first-order optical mode. The peak at 333cm⁻¹ is usually considered to be the second-order mode of E_2^{high}-E_2^{low} [12].

It is worth noting that a broad band is observed at 500-600cm⁻¹ in Mn-doped ZnO. Two peaks can be extracted from the band, which are the strong peak at 523 cm⁻¹ and the peak at 572 cm⁻¹. The peak at 523 cm⁻¹ is associated with the Mn dopant in ZnO. Hongmei Zhong et al were using a real-space recursion method to calculate the local phonon density of states (LPDOS), and the results indicated that the Mn atoms partially replaced by O atoms in the ZnO carystal lattice could cause the mode at 523 cm⁻¹ arise [13]. The peak at 572 cm⁻¹ is the longitudinal optical (LO) mode of ZnO with A_1 symmetry, which is related to intrinsic lattice defects of ZnO and often arise by the doping [12].

The intensity of the band at 500-600cm⁻¹ of 1% Mn-doped ZnO samples is much stronger than that of pure ZnO samples indicating that Mn atoms prefer to replace the O atoms in ZnO samples at this low doping concentration. Since the intensity of the band at 500-600cm⁻¹ is nearly the same in 1%, 2.5%, and 5% Mn-doped ZnO samples, the replacing of O atoms must first reach an equilibrium point, and then, when the doping concentration further increased, the Mn dopant began to enter other place in ZnO lattice.

3.3. *Band gap*

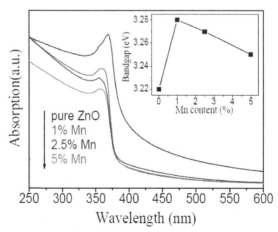

Figure 4. Absorption spectra of pure ZnO and Mn-doped ZnO dispersed in aqueous solution.. The inset shows the change of optical band gap energy due to Mn-doping.

Absorption spectra were taken to derive the width of optical band gap of Mn-doped ZnO nanorods as shown in figure 4, and the inset in figure 4 shows the change of optical band energy due to Mn-doping. The doping of Mn element can make the band gap increase, which is caused by the excessive carriers filling some energy levels on the edge of conduction band [14]. When the doping concentration of Mn increases, more Mn^{2+} ions substitute the position of Zn^{2+}, ions the strong sp-d exchange interaction causes the absorption edge shift to red side [15].

3.4. Photocatalytic activity

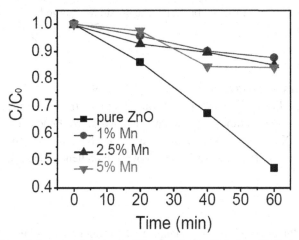

Figure 5. Degradation rates of rhodamine B by pure ZnO and Mn-doped ZnO photocatalyst.

Figure 5 shows the degradation rates of rhodamine B (RB) using pure ZnO and Mn-doped ZnO as photocatalysts. Although the increase of band gap caused by doping should enhance the photocatalytic activity of ZnO under UV light as I reported in elsewhere [16], the actual result shows the Mn-doping causes a reduction in photodegradation activity. So there must be other factors beside the change of band gap that can strongly influence the photocatalytic efficiency. As for a cationic dye, the RB is easy to attach to negatively charged object, and it could attach to neutral object as well by physcial adsorption. But the Mn-doped ZnO is positive excess, for the doping ions of Mn^{2+} tends to replace the O^{2-} in the ZnO samples prepared by our method. The RB is hard to be brought near the Mn-doped ZnO with the same kind of charges, so the degradation process is blocked.

60

4. Conclusion

The pure ZnO and Mn-doped ZnO nanorods were prepared through a chemical route. The EDX and XRD results confirmed the Mn ions have been doped into the ZnO lattice. The Raman spectra indicate that Mn dopants are likely to replace the site of O atoms in ZnO lattice, and reach a saturation point when the concentration of Mn-doping is 1%. The optical band gaps of Mn-doped ZnO samples are estimated from the absorption spectra. The result shows that band gap of ZnO increases after Mn-doping. The photocatalytic activity of ZnO samples is studied by the degradation of rhodamine B (RB), and it is found that pure ZnO samples have better photocatalytic activities than Mn-doped samples, due to the electric repulsion between the positive charged RB and positive charged ZnO.

Acknowledgments

This work was supported NSFC (51172022, 50972011), NSAF (10876001), the Research Fund of Co-construction Program from Beijing Municipal Commission of Education, the Fundamental Research Funds for the Central Universities. X. Q. Yan would like to thank the Beijing novel program (2008B19) and the Program for New Century Excellent Talents in University (NCET-09-0219).

References

1. X. M. Zhang, W. Mai, Y. Zhang, Y. Ding and Z. L. Wang, *Solid State Commun.* **149**, 293 (2009).
2. Q. Xiao, J. Zhang, C. Xiao and X. K. Tan, *Mater. Sci. Eng. B* **142**, 121 (2007).
3. R. Ullah and J. Dutta, J. Hazard. *Mater.* **156**, 194 (2008).
4. K. C. Barick, Sarika Singh, M. Aslam and D. Bahadur, *Microporous Mesop-orous Mater.* **134**, 195 (2010).
5. X. Q. Qiu, G. S. Li, X. F. Sun, L. P. Li and X. Z. Fu, *Nanotechnology* **19**, 215703 (2008).
6. Q. Wan, T. H. Wang and J. C. Zhao, *Appl. Phys. Lett.* **87**, 083105 (2005).
7. J. Bae, J. B. Han, X. M. Zhang, M. Wei, X. Duan, Y. Zhang and Z. L. Wang, *J. Phys. Chem. C* **113**, 10379 (2009).
8. J. Wang, Z. Jiang, Z. H. Zhang, Y. P. Xie, X. F. Wang, Z. Q. Xing, R. Xu and X. D. Zhang, *Ultrason. Sonochem.* **15**, 768 (2008).
9. J. Das and D. Khushalani, *J. Phys. Chem. C* **114**, 2544 (2010).

10. T. L. Phan, R. Vincent, D. Cherns, N. X. Nghia, M. H. Phan and S. C. Yu, *J. Appl. Phys.* **101**, 09H103 (2007).

11. I. Gorczyca, N. E. Christensen, E. L. P. Blancá and C. O. Rodriguez, *Phys. Rev. B* **51**, 11936 (1995).

12. T. C. Damen, S. P. S. Porto and B. Tell, *Phys. Rev.* **142**, 570 (1966).

13. H. M. Zhong, J. B. Wang, X. S. Chen, Z. F. Li, W. L. Xu and W. Lu, *J. Appl. Phys.* **99**, 103905 (2006).

14. B. Zhang, X. T. Zhang, H. C. Gong, Z. S. Wu, S. M. Zhou and Z. L. Du, *Phys. Lett. A* **372**, 2300 (2008).

15. K. Samanta, S. Dussan, R. S. Katiyar and P. Bhattacharya, *Appl. Phys. Lett.* **90**, 261903 (2007).

16. J. Zhao, L. Wang, X. Q. Yan, Y. Yang, Y. Lei, J. Zhou, Y. H. Huang, Y. S. Gu and Y. Zhang, *Mater. Res. Bull.* **46**, 1207 (2011).

STRUCTURE AND MAGNETIC PROPERTY OF NI-DOPED ZNO NANORODS

XIAOHUI ZHANG, XIAOQIN YAN[†], QINGLIANG LIAO, JING ZHAO

Department of Materials Physics and Chemistry, School of Materials Science and Engineering, University of Science and Technology Beijing, 30 Xueyuan Road, Beijing 100083, China

Ni-doped ZnO nanorods were successfully synthesized by hydrothermal method at 140 ℃ for 24 h. The Ni-doped ZnO nanorod was ~10 nm in diameter and ~200 nm in length. XRD pattern and Raman scattering indicate that the nanorods were single crystalline wurtzite structure and no secondary phases were found. The magnetic hysteresis loops show that the Ni-doped ZnO nanorods have diluted room temperature ferromagnetism and the saturation magnetization (Ms), residual magnetization (Mr) and coercive (Hc) are 0.04 Am^2/kg, 5.47×10-4 Am^2/kg, and 3.88kA/m, respectively.

1. Introduction

Diluted magnetic semiconductor (DMS) is a fresh material which has attracted great research attentions in current years due to its potential applications in spintronic devices, such as spin field-effect transistor, non-volatile memory devices and quantum computer[1~3]. According to theoretical calculations, via doping of transition metal elements, such as Cu, Mn, Ni, zinc oxide could become a kind of diluted magnetic semiconductor, whose curie temperature can be higher than the room temperature [4]. Furthermore, combining the wide band gap (3.37 eV), large excitation binding energy (60 meV), and UV lasing property with the room temperature ferromagnetism, ZnO-based DMS can be used for many magnetic-optic devices [5]. Currently, the ferromagnetism of Ni-doped ZnO nonmaterials such as films, nanorods, and nanowires arrays have been observed at room temperature [6~8]. The origin of ferromagnetism in DMS is still in debate. Some report the ferromagnetism is from the secondary phases [9] and others believe it is due to the intrinsic ferromagnetic [10].

[†] Corresponding author, e-mail: xqyan@mater.ustb.edu.cn

In this letter, we report the synthesis, structures and magnetic properties of Ni-doped ZnO nanorods. The Ni-doped nanorods were still hexagonal wurtzite structure. The room temperature ferromagnetic property was characterized by vibration sample magnetometer (VSM) and the magnetic origin was discussed.

2. Experimental

Zinc acetate dehydrate ($Zn(CH_3COO)_2 \cdot 2H_2O$, 0.1 mol/L), nickel acetate dehydrate ($Ni(CH_3COO)_2 \cdot 4H_2O$, 0.02 mol/L) and sodium hydroxide (NaOH, 0.16 mol/L) were dissolved into absolute ethanol to form a hydrothermal reaction solution, and polyethylene glycol 400(PEG 400) was added to be the surfactant. After 10 min ultrasonic dispersion, the reaction solution was transferred into a Teflon-lined stainless autoclave of 50 ml capacity. The tank was conducted in an electric oven at 140 ℃. After the reaction for 24 h, the obtained green products were rinsed by absolute ethanol and de-ionized water, respectively, for 3 times and dried at 60 ℃ in the air for 6h.

The morphologies of the products were characterized by field emission scanning electric microscopy (FE-SEM) (Zeiss, SUPRA-55) and high resolution microscopy (HREM) (JEOL, JEM-2010). Energy dispersive X-ray spectra (EDX) (Link-Inca), X-ray diffraction (XRD) (Rigaku, DMAX-RB), Raman spectra (SPEX 1403) and vibration sample magnetometer (VSM) (LAKE SHORE 7410) were designed to examine the composition, structures and magnetic properties of the Ni-doped ZnO nanorods.

3. Results and Discussions

The typical FE-SEM and TEM image of Ni-doped ZnO nanorods are shown in Fig. 1 (a) and (b). The nanorods are straight and smooth in surface. The diameter of the ZnO nanorods is ~10 nm and the length of the nanorods is ~200 nm. Fig. 1 (c), the EDX spectrum from the products, indicates a small amount Ni exists in the products and the content of Ni is determined to be 7.7 at. %.

X-ray diffraction was designed to study the structure of Ni-doped ZnO nanorods. Fig 2 shows the XRD pattern of Ni-doped ZnO nanorods. Three diffraction peaks can be observed, which indicate the ZnO wurtize strcture. Except for ZnO (100), (101) and (112) peaks, no extra diffraction peaks from pure Ni, Ni-related secondary phases or impurities were observed. It reveals that Ni substitutes Zn successfully in ZnO lattice.

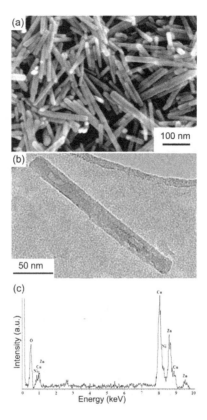

Figure 1. (a) FE-SEM morphologies of Ni-doped ZnO nanorods, (b) TEM morphology of individual Ni-doped ZnO nanorod, (c) EDX spectrum of the Ni-doped ZnO nanorod on the Cu grid

Moreover, the structure of Ni-doped ZnO nanorods was further characterized by Raman scattering. As shown in Fig 3, the Raman peaks at 98 cm^{-1} and 437 cm^{-1} denote the E_2^{low} and E_2^{high} mode of ZnO, respectively, which reveal the typical ZnO hexagonal structure. Raman peaks at 201^{-1}cm, 536cm^{-1}, and 1109cm^{-1} are classical second order modes, and they represent $2E_2^{low}$, 2LA and 2LO modes [11]. Generally, these second order Raman modes are obscure. However, in our study, the second order Raman modes is enhanced. Raman peak at 324cm^{-1} is the frequency difference between E_2^{low} and E_2^{high}, E_2^{low}-E_2^{high} [12]. Compared with those of pure ZnO, the typical Raman peaks of the Ni-doped ZnO nanorods appear red-shifted, which is related to doping effects. In addition, the Raman peak at 128 cm^{-1} is defined as TA (M) mode [13], which appears due to the zone boundary phonon at the critical point M (transverse acoustic branch at M). Nickel incorporation in ZnO induces disorder in the lattice thus breaking the crystal translational symmetry and allows zone boundary phonon to participate in Raman scattering [14].

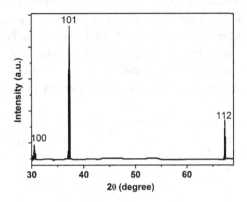

Figure 2. XRD pattern taken from Ni-doped ZnO nanorods

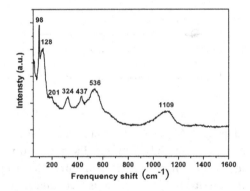

Figure 3. Raman spectrum of Ni-doped ZnO nanorods

The magnetic property of the Ni-doped ZnO nanorods was investigated by vibration sample magnetometer (VSM). Fig 4 shows the room temperature magnetic hysteresis loops of the Ni-doped ZnO nanorods. Saturation magnetization (Ms), residual magnetization (Mr) and coercive (Hc) were 0.04 Am^2/kg (0.04emu/g), 5.47×10^{-4} Am^2/kg (5.47×10^{-4} emu/g) and 3.88 kA/m (48.7Oe), respectively.

Generally, there are two controversial explanations to the origin of the room temperature ferromagnetic of Ni-doped ZnO nanorods. One is the formation of some nanoscale Ni-related secondary phases, such as metallic Ni or NiO. In this study, according to the XRD pattern and Raman scattering spectrum, there were

66

no Ni-related secondary phases existed. Moreover, NiO is antiferromagnetism at 300K [15], and Ni can be prevented during the synthesis process due to the ethanol solvent containing H_2O and OH^-. The other explanation to the ferromagnetism of Ni-doped ZnO nanorods is the result of the exchange interaction between free carriers and localized d spins on the Ni ions [5, 16]. We prefer to the latter one. The delocalized carriers are holes and electron from the valence bands. The presence of free carriers is a compulsory condition for the appearance of ferromagnetism in Ni-doped ZnO nanorods [7, 17, 18].

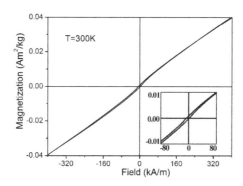

Figure 4. Magnetic hysteresis loops of the Ni-doped ZnO nanorods

4. Conclusion

In summary, Ni-doped ZnO nanorods were successfully synthesized by hydrothermal method. The nanorods, with the diameter ~10 nm and the length ~200 nm, have single crystalline wurtzite structure. The magnetic hysteresis loops show that the Ni-doped ZnO nanorods had diluted room temperature ferromagnetism. The saturation magnetization (Ms), residual magnetization (Mr) and coercive (Hc) are 0.04 Am^2/kg, 5.47×10^{-4} Am^2/kg, and 3.88kA/m, respectively. The diluted magnetism can be attributed to the exchange interaction between free carriers and localized d spins on the Ni ions.

Acknowledgements

This work was supported NSFC (51172022, 50972011), NSAF (10876001), the Research Fund of Co-construction Program from Beijing Municipal Commission of Education, the Fundamental Research Funds for the Central Universities. X. Q. Yan would like to thank the Beijing novel program

(2008B19) and the Program for New Century Excellent Talents in University (NCET-09-0219).

References

1. S. D. Sarma, Nature Mater. 2, 292 (2003).
2. S. J. Pearton, C. R. Abernethy, M. E. Overberg, G. T. Thaler, A. F. Hebard, Y. D. Park, F. Ren, J. Kim, and L. A. Boatner, H. Appl. Phys. 1, 93 (2003).
3. X. M. Zhang, Y. Zhang, Z. L. Wang, W. Mai, Y. Gu, W. Chu, and Z. Wu, Appl. Phys. Lett. 92, 162102 (2008).
4. T. Dietl, H. Ohno, F. Matsukura, J. Cibert, and D. Ferrand, Science 287, 1019 (2000).
5. C. Cheng, G. Xu, H. Zhang, and Y. Luo. Mater. Lett. 62, 1617 (2008).
6. M. Venkatesam, C. B. Figeralde, J. G. Lunney, and J. Coey, Phys. Rev. Lett. 93, 177206 (2004).
7. X. Liu, F. Lin, L. Sun, W. Cheng, X. Ma, and W. Shi, Appl. Phys. Lett. 88, 062508 (2006).
8. J. B. Cui and U. J. Gibson, Appl. Phys. Lett. 87, 133108 (2005).
9. S. Ramachandran, A. Tiwari, and J. Narayan, Appl. Phys. Lett. 84, 5255 (2004).
10. H. Wang, Y. Chen, H. B. Wang, C. Zhang, F. J. Yang, J. X. Duan, C. P. Yang, Y. M. Xu, M. J. Zhou, and Q. Li, Appl. Phys. Lett. 90, 052505 (2007).
11. R. Cuscó, E. A. Lladó, J. Ibáñez, L. Artús, J. Jiménez, B. Wang, and M. J. Callahan, Phys. Rev. B. 75, 165202 (2007).
12. B. B. Li, X. Q. Xiu, R. Zhang, Z. K. Tao, L. Chen, Z. L. Xie, Y. D. Zheng, and Z. Xie, Mater. Sci Semicon. Proc. 9, 141 (2006).
13. H. K. Yadav, K. Sreerivas, V. Gupta, J. F. Scott, and R. S. Katiyar, Appl. Phys. Lett. 92, 122908 (2008).
14. H. K. Yadav, K. Sreerivas, V. Gupta, R. and S. Katiyar, J Appl. Phys. 104, 053507 (2008).
15. D. A. Shcwartz, K. R. Dittlstved, and D. R. Gamelin, Appl. Phys. Lett. 85, 1395 (2004).
16. K. R. Kittlstved, W. K. Liu, and D. R. Gamelin, Nature Matt. 5, 291 (2006).
17. D. Wu, M. Yang, Z. Huang, G. Yin, X. Liao, Y. Kang, X. Chen, and H. Wang. J Coll. Interface Sci. 330, 380 (2009).
18. J. Zhao, L. Wang, X. Yan, Y. Yang, Y. Lei, J. Zhou, Y. Huang, Y. Gu, and Y. Zhang, Mater. Res.Bul. 46, 1207(2011).

DOPING EFFECT OF THE ELECTRONIC TRANSPORT PROPERTIES OF ZINC OXIDES NANOWIRES STUDIED BY FIRST PRINCIPLES CALCULATION

YOUSONG GU[‡], XU SUN, XUEQIANG WANG AND YUE ZHANG

Department of Material Physics and Chemistry, University of Science and Technology Beijing, Beijing 100083, People's Republic of China

First principles calculations were performed to study the electronic structures and electronic transport properties of both p-type and n-type ZnO nanowries via the Siesta/Transiesta codes. It is found that Li, Na and K doped ZnO nanowires are typical p-type semiconductors and show good linear I ~ V characteristics at low bias. The range for linear I ~ V characteristics decreases as the dopant change for Li to K. Al, Ga doped ZnO nanowires are typical n-type semiconductors with half-filled conduction bands and show good metallic I ~ V characteristics in a large range. In the case of In doped ZnO nanowires, half-filled impurity band of In atom lies in the band gap, and linear I ~ V curve can be seen only at small bias.

1. Introduction

It is well known that zinc oxide is a direct and wide band gap (E_g=3.37eV) semiconductor with high exciton binding energy (E_b=60meV). Due to the interesting electronic, piezoelectric and photoelectric properties, ZnO nanowires have a wide range of applications in electronic, piezo-electronic, optoelectronic devices and sensors [1-5]. The electronic transport properties of ZnO nanowires (NWs) are extremely important to the design and optimization of nanodevices [6].

Many experimental results were available on the electronic transport properties of ZnO NWs [7]. Electronic transport properties of doped ZnO nanowires under strains were also reported [8, 9]. However, only a few theoretical works on the transport properties of ZnO nanowires were reported [10-12]. Kamiya et al. [10] calculated the electronic transport properties of bulk ZnO coupled by Au or Mg electrodes and found Schottky and Ohmic contact at the Au/ZnO/Au and Mg/ZnO/Mg interfaces, respectively. Yang et al. [11] studied the electronic transport properties of ZnO nanowires coupled by

---

[‡] Corresponding author, e-mail: yousongu@mater.ustb.edu.cn

aluminum electrodes and observed clearly rectifying I ~ V characteristics. The length dependence of electronic transport for single-walled ZnO nanotubes was investigated by Qin Han et. al. [12] and found that the conductance decreased exponentially with the length of nanotubes at low bias but the current were insensitive to the lengths at high bias.

Doping is often used to optimize the properties of nanomaterials for device application. However, theoretical studies on the electronic transport properties of doped ZnO nanowires have not been reported in literatures yet.

In this work, first principles calculations based on density functional theory and non-equilibrium Green function (NEGF) were performed to study the electronic structure and transport properties of ZnO nanowires with both n-type and p-type doping.

2. Computational Detail

First-principles calculations were performed in the frame work of density functional theory and non-equilibrium Green function using the Siesta/Transiesta package [13-14], which employed pseudopotentials and numerical atomic orbit basis sets. In the calculation, generalized gradient approximations in the form proposed by Perdew, Burke, and M. Ernzerhof (GGA-PBE) were chosen as the exchange correlation functionals [15], and the size of the basis sets was double zeta polarized (DZP). The mesh cutoff was 250 Ry and 1x1x30 grid were used for k-point sampling in the calculation of electronic structures. The supercell for nanowire was chosen as a big box with one unit of nanowire located in the center with plenty of spacing around so that the c-axis length was just a period of the nanowire and the separation between the nanowires was larger that 1.0 nm. Doping was realized by replace one Zn atom with a dopant. Geometry relaxation were performed before electronic structure calculation. The band structure and local density of state near the Fermi level were obtained.

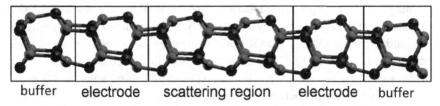

buffer   electrode   scattering region   electrode   buffer

Figure 1. The atomic model used to calculate the transport properties of ZnO nanowires.

In order to calculate the electronic transport properties of ZnO nanowires, 1x1x3 supercells were built and different regions were designated in the sequence of left buffer, left electrode, scattering region, right electrode and right buffer, as shown in Figure 1. Transiesta calculation was performed for the electrode first, and then the whole system was calculated to get the Hamiltonian and overlapping of the system (TSHS file) at each bias voltage. Transmission spectra and I ~ V curve were obtained by the tbtran post processing tool.

3. Results and Discussion

3.1. *Li, Na and K doped ZnO nanowires*

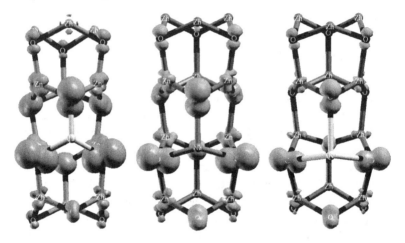

Figure 2. The local density of states (LDOS) near the Fermi Level for Li, Na and K doped ZnO NWs.

A Zn atom in a supercell was substituted by Li, Na or K atom to make p-type doping. The electronic structures of the p-type doped ZnO nanowires were calculated after full geometry relation by the SIESTA code. Figure 2 shows the atomic positions after geometry relaxation and the local density of states (LDOS) near the Fermi level as shown by the iso-surfaces (isovalue=0.01 states/Bohr^3). The lattice parameters and atomic positions change as the radii of the dopant changes. It is clear that LDOS near Fermi level is distributed all over the atoms in the supercells due the charge transfer caused by doping with Li, Na or K atoms. Good conduction behaviors are expected.

The detailed band structures at the top of the valence bands of the doped ZnO nanowires are shown in Figure 3. The Fermi levels are located at the middle of the highest valence bands and it is a typical p-type semi-conductor.

The width of the top valence bands decrease from Li-doped to K-doped ZnO nanowires.

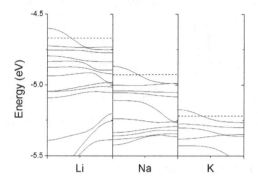

Figure 3. The band structures of Li, Na and K doped ZnO nanowires. The dotted lines are the Fermi level.

The electronic transport properties of ZnO nanowires were calculated by the Transiesta code. Ballistic transmission regions can be seen in the center of the transmission spectra as shown in Figure 4. Electrons can pass through without any resistance in the black rhombic regions. However, electrons can't pass through in the forbidden white area. It is also found that the area of the rhombic region decreases as the dopant changes from Li to K. This is in agreement with the band structure.

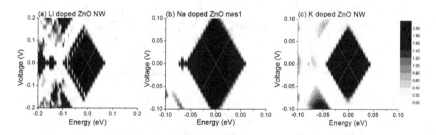

Figure 4. The transmission spectra of Li, Na and K doped ZnO NWs.

The calculated I ~ V characteristic curves are shown in Figure 5. It can be seen that linear I ~ V relationship are observed in low bias regions as expected. However, I ~ V curves deviates from linear relationship as the bias increases. This can be explained by the transmission spectra and that is closely related to the detailed band structures. Therefore, good conductivity and linear I ~ V curves are observed in Li, Na and K doped ZnO nanowires at low bias and the range for linear I ~ V curve decreases as the dopant change for Li to K.

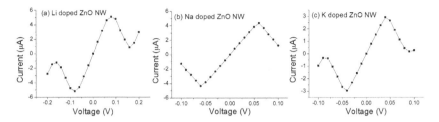

Figure 5. The I ~ V curves of Li, Na and K doped ZnO NWs. (a) Li doped (b) Na doped (d) K doped ZnO nanowires.

3.2. Al, Ga and In doped ZnO nanowires

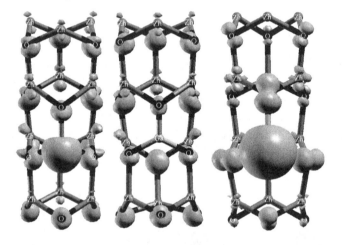

Figure 6. The local density of states (LDOS) at Fermi Level for Li, Na and K doped ZnO.

A Zn atom in a supercell was substituted by Al, Ga or In atom to make n-type doping. The electronic structures of the n-type doped ZnO nanowires were calculated after full geometry relation by the SIESTA code. Figure 6 shows the atomic positions after geometry relaxation and the local density of states (LDOS) near the Fermi level as shown by the iso-surface (isovalue=0.001 states/Bohr^3 for Al, Ga doping, and isovalue=0.005 states/Bohr^3 for In doping). The lattice parameters and atomic position change as the radus of the dopant changes. It is clear that LDOS near the Fermi level is distributed all over the atoms in the supercells due the charge transfer by doping with Al or Ga. Good, metallic conductances are expected. In the case of In doping, large amount of states are located near the In atom, indicating that the Fermi level lies at the impurity band of In atoms in the supercell.

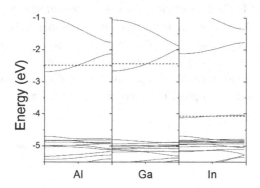

Figure 7. The band structure of Al, Ga and In doped ZnO nanowires.

The band structures of the n-type doped ZnO nanowires are shown in Figure 7. In case of Al and Ga doped ZnO nanowires, the Fermi levels are located at the middle of the lowest conduction bands and it is typical p-type semi-conductors. The widths of the conduction bands are much wider than the top valance band in the Li, Na and K doped ones. However, in the case of In doped ZnO nanowire, the Fermi level lie at the localized impurity band in the middle of the band gap. Good conduction is also expected since it is only half filled.

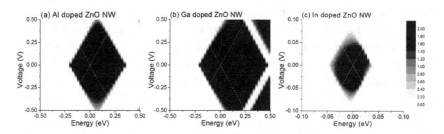

Figure 8. The transmission spectra of Al, Ga and In doped ZnO nanowires.

Large ballistic transmission regions can be seen in the center of the transmission spectra of Al and Ga doped ZnO nanowires, as shown in Figure 8. Electrons can pass through without any resistance in the black rhombic regions,

74

since the transmission coefficient is nearly 2.0 (for two spins). The large rhombic regions correspond to the wide conduction bands in Al and Ga doped ones. In the case of In doping, a small rhombic region is observed, which corresponds to the narrow impurity band.

The calculated I ~ V characteristic curves are shown in Figure 9. It can be seen that linear I ~ V relationship is observed in a large region corresponding to the large rhombic region for Al and Ga doping as expected, and linear I ~ V relationship is observed only in small region corresponding to the small rhombic region. Therefore, good conductivity and linear I ~ V curves are observed in wide range in Al and Ga doped ZnO nanowires and only in small range in In doped ZnO nanowires.

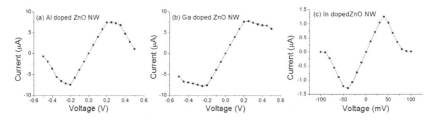

Figure 9. The I ~ V curves of Li, Na and K doped ZnO NWs.

4. Conclusions

The electronic structures and transport properties for both p-type and n-type ZnO nanowires were studied by the Siesta/Transiesta codes. It is found that Li, Na and K doped ZnO nanowires are typical p-type semiconductors and show good linear I ~ V characteristics at low bias. The range for linear I ~ V characteristics decreases as the dopant change for Li to K. Al, Ga doped ZnO nanowires are typical n-type semiconductors with half-filled conduction bands and show good metallic I ~ V characteristics in large ranges. In the case of In doped ZnO nanowires, half-filled impurity band located at In atom lies in the band gap, and linear I ~ V curve can be seen at small bias.

Acknowledgments

This work was supported by the Major Project of International Cooperation and Exchanges (2006DFB51000), NSFC (50972009), NSAF (10876001), the Research Fund of Co-construction Program from Beijing Municipal Commission of Education, the Fundamental Research Funds for the Central Universities.

References

1. Y. Yang, J. J. Qi, W. Guo, Y. S. Gu, Y. H. Huang and Y. Zhang, *Phys. Chem. Chem. Phys.*, **12**, 12415–12419 (2010).

2. X. M. Zhang, M. Y. Lu, Y. Zhang, L. J. Chen, and Z. L. Wang, *Adv. Mater.*, **21**, 2767–2770 (2009).

3. Y. Yang, J. J. Qi, Q. L. Liao, H. F. Li, Y. S. Wang, L. D. Tang and Y. Zhang, *Nanotechnology*, **20**, 125201 (2009).

4. Z. L. Wang and J. H. Song, *Science*, **312**, 242 (2006).

5. Y. Lei, X. Q. Yana, N. Luo, Y. Song, Y. Zhang, *Colloids and Surfaces A: Physicochem. Eng. Aspects*, **361**, 169–173 (2010).

6. Y. Hu, Y. Zhang, C. Xu, G. Zhu, and Z.L. Wang, Nano Lett. **10**, 5025(2010).

7. P.-X. Gao, Y. Ding and Z.L. Wang, Nano Lett. **9**, 137 (2009).

8. Y. Yang, J.J. Qi, Y. Zhang, Appl. Phys. Lett. **92**, 182117 (2008).

9. K.H. Liu, P. Gao, Z. Xu, Appl. Phys. Let. **92**, 213105 (2008).

10. T. Kamiya, K. Tajima, K. Nomura, H. Yanagi, and H. Hosono, *Phys. Status Solidi A*, **205**, 1929 (2008).

11. Z. J. Yang, L. H. Wan, Y. J. Yu, Y. D. Wei and J. Wang, *J. Appl. Phys,.* **108**, 033704 (2010).

12. Q. Han, B. Cao, L. P. Zhou, G. J. Zhang, and Z. H. Liu, *J. Phys. Chem. C*, **115**, 3447–3452 (2011).

13. E. Artacho, E. Anglada, O. Dieguez, J. D. Gale, A. García, J. Junquera, R. M. Martin, P. Ordejón, J. M. Pruneda, D. Sánchez-Portal and J. M. Soler, J. Phys.: Condens. Matter, **20**, 064208 (2008).

14. J.M. Soler, E. Artacho, J.D. Gale, A. García, J. Junquera, P. Ordejón, and D. Sánchez-Portal, J. Phys. Condens. Matter, **14**, 2745 (2002)

15. J.P. Perdew, K. Burke, and M. Ernzerhof, Phys. Rev. Lett., **77**, 3865 (1996)

ELASTIC RESPONSE OF COPPER SLABS TO BIAXIAL STRAIN STUDIED BY DENSITY FUNCTIONAL THEORY CALCULATION

XUEQIANG WANG, YOUSONG GU[§] AND XU SUN

Department of Material Physics and Chemistry, University of Science and Technology Beijing, Beijing 100083, People's Republic of China

Nanomaterials have different elastic properties from that of bulk, as revealed by many experimental and theoretical investigations. The elastic energy of a biaxially strained material depends on both the magnitude and the direction of the applied biaxial strain. In application to thin homogeneously strained slabs, these results enable us to estimate the different response between the outmost layer and bulk layers. The Poisson ratio and elastic energy of copper slabs under biaxial strain in (111) and (001) direction are studied by density-functional theory calculations. For (111) models the biaxial Poisson ratios of outmost layers are 30% larger than that of the bulk, and for (001) models the changes is smaller, about 7%. It can be seen that the changes in biaxial Poisson ratios depends on the orientation and no dependence on thickness is observed.

1. Introduction

During last decades, surface science has focused on surfaces of macro-scopic single crystal samples. In such surfaces samples, there would be strained surface atoms and bulk substrate, which is strain-free. The unstrained bulk substrate influences properties of the surface layers, but the strained surface has little or no effect on the substrate properties. For the history above, the measurement, the definition, and calculation of these quantities for surfaces science were based on these two assumptions, strained surface atoms and strain-free substrates.

Recently, with the development of nanotechnology, some surface techniques and concepts are being used to nanoscale objects. Objects at this scale have different elastic properties from that of bulk, as revealed by many experimental and theoretical investigations [1-4]. In such case, the surface atoms will take large effects on interior region than the sample we used for decades. The basic assumptions about strain-free substrates should be changed. We call the interior region "bulk part" in this paper.

[§] Corresponding author: e-mail: yousongu@mater.ustb.edu.cn

According to this idea, we study on the elastic response of copper slab model with biaxial strain. As a matter of fact, nano-sheet is not a true 2D system. The slab models are realistic 3D models and present first-principal simulations of the energy of Cu (111) and (001) sheets with different thickness as a function of the in-plane strain components. It does not make any sense apply stress in perpendicular direction, but when we apply stress in parallel direction, the slab model will have a strain in perpendicular to reduce the total energy of the system. How this part would responds? Will it be different from the bulk? In this paper, we will give an answer to such question.

2. Elastic response of cubic crystals to biaxial strain

Poisson's Ratio is the ratio of transverse contraction strain to longitudinal extension strain in the direction of stretching force (Fig 1). The general expressions for uniaxial Poisson's ratio v(hkl) along arbitrary loading direction [hkl] are given for cubic crystals by J. Zhang et al.[5]. Take (100) surface as an example, the Poisson's ratio v can be expressed as

$$v = C_{12} / (C_{11} + C_{12})$$ (1)

We show this uniaxial Poisson's ration to compare with the Biaxial one's which will be mentioned below. With different strais, we will have different results.

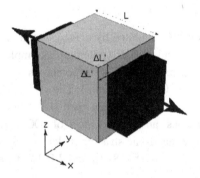

Figure 1. A cube with sides of length L of an isotropic linearly elastic material subject to tension along the x axis, with a Poisson's ratio of 0.5. The green cube is unstrained, the red is expanded in the x direction by ΔL due to tension, and contracted in the y and z directions by ΔL'.

For the case in slab model or free standing thin film, we will stretch the model with in plane equi-biaxial strain. T. Hammerschmidt[6] derivate the general analytic expressions of the strain tensor, the Poisson ratio, and the elastic energy for cubic systems under biaxial strain within linear-response. The biaxial strain tensor in this coordinate system of the deformation is

$$\varepsilon = \begin{pmatrix} \delta & 0 & 0 \\ 0 & \delta & 0 \\ 0 & 0 & -v\delta \end{pmatrix} \tag{2}$$

where v is the biaxial Poisson ratio. This biaxial Poisson ratio depends on the elastic constants c_{ij} and the strain plane (hkl):

$$v = 2\frac{c_{12}(h^4 + k^4 + l^4) + (c_{11} + c_{12} - 2c_{44})(h^2k^2 + h^2l^2 + k^2l^2)}{c_{11}(h^4 + k^4 + l^4) + 2(c_{12} + 2C_{44})(h^2k^2 + h^2l^2 + k^2l^2)} \tag{3}$$

For (001) and (111) plane, the biaxial Poisson ratio of bulk material is

$$v_{(001)} = 2\frac{C_{12}}{C_{11}} \tag{4}$$

$$v_{(111)} = 2\frac{C_{12} + (C_{11} + C_{12} - 2C_{44})}{C_{11} + 2(C_{12} + 2C_{44})} \tag{5}$$

3. Calculation method

The elastic properties were calculated by density-functional theory (DFT) using a plane-wave basis set and pseudo potential as implemented in VASP[7]. Generalized gradient approximation (GGA) potentials were used. We calculated from 5 layers to 17 layers (0.7nm to 2.9nm). The slab model was simulated as a periodic unit cell with a vacuum between each slab at least 15Å thick. In these bulk and slabs calculations, an energy cutoff of 400eV was used to get more convergence results. The electronic structure calculations were carried out using Monkhorst-Pack grid of up to 18×18×18 and 14×14×3 for the bulk and slabs, respectively.

4. Results and analysis

4.1. *Bulk calculation*

The internal energy convergence with energy cutoff and the k-point grid size was tested, starting from 225eV and 4×4×4 k-point grid respectively. The internal energy converged at meV level after 375eV with 11×11×11 k-points grid. The main idea of elastic constants calculation is that treat the total energy difference as a polynomial function of the strain, select specific deformation modes to reduce the number of the strain components that will appear in the internal energy function. All these calculation should near the lattice equilibrium position. We use Wang and Ye[8]'s method to calculate the bulk elastic constants listed in table 1. The biaxial Poisson ratio is also calculated use both calculated results and experimental results using the expression (1), (4) and (5) respectively. These calculation results are in good agreement with the experimental results.

Table 1 Comparison with calculation and experimental results

	Calculation Results (400,14*14*14)(GPa)	Exp Results(GPa)[a]
C11	176.2	169
C12	121.0	122
C44	72.9	75.3
v(bulk)[b]	0.407	0.42
v(001)[c]	1.373	1.44
v(111)[d]	0.767	0.73

Reference a [9]
Value b, c and d are calculated with expressions(1),(4) and (5) respectively.

4.2. Slab models calculation

Table 2 Relaxed lattice parameter a for (111) and (001) slab models

	(111) slab (Å)	(001) slab (Å)
5 Lyers	2.547	3.576
7 Lyers	2.557	3.587
9 Lyers	2.562	3.599
11 Lyers	2.565	3.603
13 Lyers	2.567	3.608
15 Lyers	2.568	3.610
17 Lyers	2.569	3.621
Bulk	2.577	3.644

To get an accurate description for the surface energy, we need to build slabs with 5 to 17 Cu(111) and Cu(100) mono-layers. All slabs are separated by 15Å thick vacuum space. As a slab model, the in-plane lattice constant and the interlayer distances are allowed to relax to their equilibrium values. The system will not be equilibrium until it reaches the minima total energy. The relaxed lattice parameters are listed in table 2. We can see that the equilibrium lattice parameter is smaller when the slab gets thinner.

The distances between adjacent layers are shown in figure 2. The outmost layers' spacing is smaller than the layer spacing of the bulk part. The inter-planar spacing tends to converge on the bulk value, with the increase of the layer number. T M Trimble and R C Cammarata mentioned such result in recent literature[10]. Comparing with (111) slab, (100) slab has a larger amount of change in layer spacing.

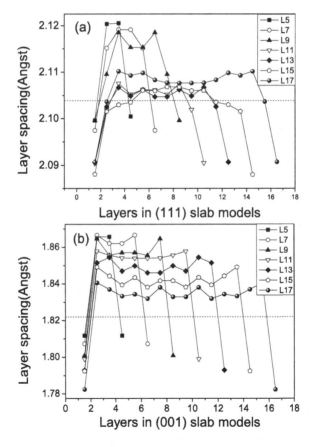

Figure 2. The distance between adjacent layers for (a) Cu (111) and (b) Cu(001) slab models. Compare with the (111) surface, (100) has a larger interlayer spacing change.

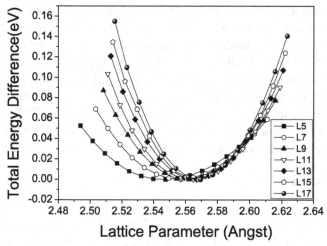

Figure 3. Total energy difference for (111) slab models. The equilibrium lattice parameter is smaller when the slab gets thinner.

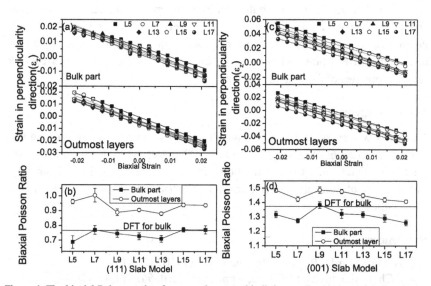

Figure 4. The biaxial Poisson ratio of outmost layers and bulk layers. For (111) models the biaxial Poisson ratios of outmost layers are 30% larger than that of the bulk, and for (001) models the changes is smaller, about 7%. It can be seen that the changes in biaxial Poisson ratios depends on the orientation and no dependence on thickness are observed.

After relaxation, we apply a series of in-plane biaxial strain on the slab models from -0.021 to 0.021 with step 0.003. The Energy-strain cuvres are shown in Figure 3. We analyse the strain in perpendicularity direction of bulk part and outmost layers, and use the expression (2) to generate the biaxial Poisson ratio for both cases, and the results are shown in Figure 4. We found that the biaxial Poisson ratios of the bulk part are in good agreement with those of continuum-elasticity theory. Especially for (111) slab model since it is the most stable surface of copper. For (111) models the biaxial Poisson ratios of outmost layers are 30% larger than that of the bulk, and for (001) models the changes is smaller, about 7%. It can be seen that the changes in biaxial Poisson ratios depends on the orientation and no dependence on thickness are observed.

5. Conclusion

Thin slab models were used to estimate the biaxial Poisson ratio of freestanding copper films at several nanometers scales. The biaxial Poisson ratios of bulk part are the same as those of continuum-elasticity theory, especially for the most stable (111) slab models. On the contrary, the outmost layers have different elastic responses. For (111) models the biaxial Poisson ratios of outmost layers are 30% larger than that of the bulk, and for (001) models the changes is smaller, about 7%. The variation of different slab model series shows that, the biaxial Poisson ratios of outmost layers are only orientation depends. From the above results, we have a better understanding on the size and orientation effects on elastic properties of free standing slabs.

Acknowledgement

This work was supported by the Major Project of International Cooperation and Exchanges (2006DFB51000), NSFC (50972009), NSAF(10876001), the Research Fund of Co-construction Program from Beijing Municipal Commission of Education, the Fundamental Research Funds for the Central Universities.

References:

1. M. Durandurdu, Physical Review B **81**, 174107 (2010).
2. J. Choi, M. Cho, and W. Kim, Appl. Phys. Lett. **97**, 171901 (2010).
3. R. Cherian *et al.*, Physical Review B **82**, 235321 (2010).
4. J. C. Hamilton, and W. G. Wolfer, Surf. Sci. **603**, 1284 (2009).
5. J. Zhang *et al.*, J. Phys. Chem. Solids **68**, 503 (2007).
6. T. Hammerschmidt, P. Kratzer, and M. Scheffler, Physical Review B **75**, 235328 (2007).

7. G. Kresse, and J. Furthmüller, Physical review. B, Condensed matter **54**, 11169 (1996).
8. S. Q. Wang, and H. Q. Ye, J. Phys.: Condens. Matter **15**, 5307 (2003).
9. H. M. Ledbetter, J. Phys. Chem. Ref. Data **3**, 897 (1974).
10. T. M. Trimble, and R. C. Cammarata, Surf. Sci. **602**, 2339 (2008).

ELECTRONIC TRANSPORT PROPERTIES OF ONE DIMENSIONAL ZNO NANOWIRES STUDIED USING MAXIMALLY-LOCALIZED WANNIER FUNCTIONS

XU SUN, YOUSONG GU*, XUEQIANG WANG

School of Materials Science and Engineering, University of Science and Technology Beijing, 100083, Beijing, P R China

One dimensional ZnO NWs with different diameters and lengths have been investigated using density functional theory (DFT) and Maximally Localized Wannier Functions (MLWFs). It is found that ZnO NWs are direct band gap semiconductors and there exist a turn on voltage for observable current. ZnO nanowires with different diameters and lengths show distinctive turn-on voltage thresholds in I-V characteristics curves. The diameters of ZnO NWs are greatly influent the transport properties of ZnO NWs. For the ZnO NW with large diameter that has more states and higher transmission coefficients leads to narrow band gap and low turn on voltage. In the case of thinner diameters, the length of ZnO NW can effects the electron tunneling and longer supercell lead to higher turn on voltage.

1. Introduction

Owing to the unique electronic, optical, piezoelectric properties, ZnO nanostructures have been extensively investigated for the applications in field effect transistors [1], light-emitting diodes [2], biosensors [3], and piezoelectric devices [4], and nanogenerators [5]. However, profound understanding of behavior of charge carriers in ZnO nanostructures is still a challenge and need to be solved to facilitate the designs and fabrications of high performance ZnO-based devices as well as understand the fundamental scientific problems.

In the decade, there are huge experimental studies on the electronic transport properties [6-10], but many uncontrollable factors limited the development of ZnO-based nanostructures to meet the practical applications. Although first-principles calculations based on density functional theory are well-established and widely used for simulating the physical properties of ZnO, the studies of the transport properties of ZnO nanowires and nanobelts were few, and only began to be carried out in the last few years [11-13]. Recently, Kamiya

---

* Corresponding author, e-mail: yousongu@mater.ustb.edu.cn

et al. [11] calculated the carrier transport of bulk ZnO coupled by Au and Mg electrodes. They found that the Au/ZnO/Au interface showed Schottky contact behavior while the Mg/ZnO/Mg interface showed Ohmic contact. Yang et al. [12] studied electronic transport properties of ZnO nanowires coupled by aluminum electrodes and observed clearly rectifying current-voltage characteristics. The electrical transport properties for the length dependence of single-walled ZnO nanotubes were investigated by Qin Han et al. [13].

In this article, in order to reveal the intrinsic charge carrier of ZnO NWs, we concentrated on the transport behaviors of one dimensional ZnO NWs with homo-electrodes neglecting the interface between electrodes and central scattering region, to reveal the intrinsic carrier properties of ZnO nanowires using the density functional theory (DFT) and Maximally Localized Wannier Functions (MLWFs) [14], which could help us distinguish the distribution of the transport properties of ZnO NWs from the contact or interface between the ZnO NWs with mental electrodes.

2. Calculation Methods

Our calculations were performed with the Quantum-ESPRESSO package [15] with ultrasoft Vanderbilt pseudopotentials [16]. Wave functions of nanostructured systems used to calculate quantum conductance were computed using Wannier90 which was included in the Quantum-ESPRESSO package. The generalized gradient approximation (GGA) in the scheme of Perdew-Wang 91 was adopted. DFT combined with the Landauer formulation [17] has become a standard starting point for evaluating quantum conductance.

To calculate the electronic transport, first of all, self-consistent calculation using Quantum-ESPRESSO are performed; as the delocalized Bloch orbitals cannot be directly used to calculate transport properties with the method mentioned above, then we need the matrix elements of the Hamiltonian calculated in a localized-orbital basis. The overlap between Bloch states and the projections are obtained. Next, disentangle and wannierise procedure can then be used to minimize self-energy and hence obtain MLWF for this optimal subspace. The electronic current can be obtained by Landauer-Büttiker formula:

$$I = \frac{2e^2}{h} \int T(E)[f_L(E, \mu_L) - f_R(E, \mu_R)] dE \qquad (1)$$

where T(E) is the transmission matrix and is obtained by Green's function via the following formula:

$$T(E) = \Gamma_L G^r \Gamma_R G^a$$

$$f_{L/R} = 1/\{1 + \exp[(E - \mu_{L/R})/kT]\}^{-1} \qquad (2)$$

is the Fermi-Dirac distribution function of left/right electron reservoir connected to the left/right electrode, $\mu_{L/R}$ is the electrochemical potentials of left/right lead.

In this work, we focus on the internal charge carriers in the nanowires by select the same ZnO nanowires as electrodes. In our calculation, the kinetic energy cut-off for wavefunctions is 30 Ry (~ 400 eV), a 1 × 1 × 10 mesh of k-points is used to sample the Brillouin zone. Wave function projections were calculated for Zn l=0, 2 and O l=1 corresponding to the valence manifolds and several lower conduct manifolds.

3. Results and Discussions

In order to study the effect of diameter and length of ZnO Nanowires (ZnO NWs) on the conductance, we have built ZnO NWs with diameters of 4.64 and 7.74 Å as represented in Figure 1(a) and (b), and named as ZnO-NWn indicating that the cross section is made of n hexagons. Length effect has been studied by change the periods of ZnO NWs and named by adding a suffix –z1 or -z2, as shown in Figure 1(c) and (d).

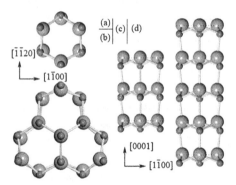

Figure 1. Structural schemes of ZnO Nanowires with small (a) and large diameter (b) in the top view; and with single (c) and double periods (d) from side view. big balls and small balls represent Zn and O atoms respectively.

Both the atomic positions and supercell parameters were fully relaxed; individual forces are less than 0.003 Ry/au. It is found that Zn-O bonds in c direction (1.89 Å) was shorter than that of the bulk (1.99 Å), while the bond in the cross-section is similar to that in bulk (1.97 Å) and slightly tilted. O atom relaxed outward for about 0.3 Å. The structural and electronic properties are in good agreement with previous DFT calculations.

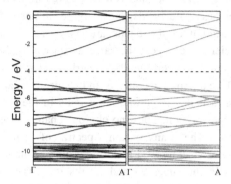

Figure 2. Band structure of DFT (black line) and Interpolated band structures with Wannier functions (gray line). The dash dot line in the middle of graph is the Fermi level.

In order to evaluate the charge carriers with MLWFs basis set, the quality of the disentanglement may be assessed by comparing the interpolated band structure provided by Wannier90 to the full band structure given by the DFT code. The match between the Wannier interpolation and the full band structure is excellent. We see that the interpolated band structure in Figure 2 describes perfectly conduction states up to about 2.5 eV above the Fermi energy.

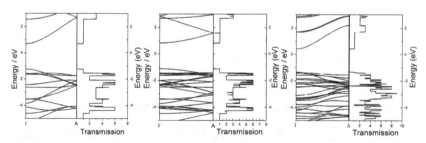

Figure 3. Band structure and transmission spectra for (a) ZnO-NW1-z1, (b) ZnO-NW1-z2, and (c) ZnO-NW3-z1.

The band structures and transmission spectra of three systems were shown in Figure 3. Fermi Levels in the middle of the gap are set to zero. According to the band structure of ZnO-NWs in Figure 3, the highest occupied electronic states of valence bands near the Fermi energy are mainly contributed by the O 2p-orbitals, and deeper part of valence bands are hybridized by Zn 3d-orbitals and O 2p-orbitals. For comparison, the conduction bands are contributed by the hybridized states of Zn 4s/4p-oribitals and O 2p-orbitals. The transmission coefficient steps along with the number of bands, and there is no transmission in the gap. Comparing the case of ZnO-NW1-z1 and ZnO-NW1-z2, it could be found that the later has more bands due to large number of electrons in the supercell but the location of the bands are the same. It is also found that transmission coefficients are almost the same as shown by the positions of the steps since the transmission coefficient is directly related to the band structure. As there is no interface among the contacts and electrodes and the charge carrier inside the nanowire is ballistic, periodic unit cells in the supercell should not affect the electron transmission. Thus, no obvious change could be found in the transmission spectra. The transmission spectra of ZnO-NW3-z1 possess different characteristics as compared to formers. Larger diameter leads to more electrons and states in the supercell and results in denser bands in the band structure as shown in Figure 3(c). The transmission spectrum calculated by Wannier90 is shown in Figure 3(c) and dense and high peaks were found in the transmission spectrum due to large number of states and carrier channels.

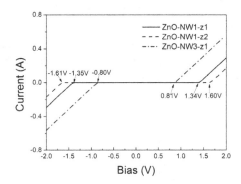

Figure 4. I ~ V curves for the three ZnO nanowires

The I ~ V curves of the three ZnO NWs are calculated by the Landauer-Büttiker formula expressed by Eq. (1), as indicated by the lines with different style in Figure 4. The obvious Schottky type I-V characteristic curves were obtained and the turn-on voltage of ZnO NW is 1.34V, 1.60 V and 0.81 V for three ZnO nanowires, respectively. The change of band gap determines the I-V characteristics. As the band gaps of ZnO NWs are gradually decreased as the diameter of ZnO NWs increases, the ZnO NW with larger diameter owns lower turn-on voltage. Comparing the case of ZnO-NW1-z1 and ZnO-NW1-z2, it shows that long supercell has large turn-on voltage. The increase of the supercell length elongates the distance for electron tunneling, which leads to increased difficulty for electron tunneling and increased the turn-on voltage for the ZnO NWs.

4. Summary

The electron structures and transport properties of the ZnO NWs for different length and diameters were studied via MLWF method. It is found that ZnO NWs are direct band gap semiconductors and there exist a turn on voltage for observable current. The diameters of ZnO NWs can greatly influent the electronic structure and transport property of ZnO NWs. ZnO NW with larger diameter has more states, higher transmission coefficients, narrower band gap and lower turn on voltage. The length of ZnO NW can also effects electron tunneling and longer supercell lead to higher turn on voltage.

Acknowledgments

This work was supported by the Major Project of International Cooperation and Exchanges (2006DFB51000), NSFC (50972009), NSAF (10876001), the Research Fund of Co-construction Program from Beijing Municipal Commission of Education, the Fundamental Research Funds for the Central Universities.

References

1. Y. Yang, J. J. Qi, W. Guo, Y. S. Gu, Y. H. Huang and Y. Zhang, *Phys. Chem. Chem. Phys.*, **12**, 12415–12419 (2010).
2. X. M. Zhang, M. Y. Lu, Y. Zhang, L. J. Chen, and Z. L. Wang, *Adv. Mater.*, **21**, 2767–2770 (2009).
3. Y. Lei, X. Q. Yana, N. Luo, Y. Song, Y. Zhang, *Colloids and Surfaces A: Physicochem. Eng. Aspects*, **361,** 169–173 (2010).
4. Y. Yang, J. J. Qi, Q. L. Liao, H. F. Li, Y. S. Wang, L. D. Tang and Y. Zhang, *Nanotechnology*, **20**, 125201 (2009).
5. Z. L. Wang and J. H. Song, *Science*, **312**, 242 (2006).
6. Y. W. Heo, L. C. Tien, and D. P. Norton, B. S. Kang and F. Ren, B. P. Gila and S. J. Pearton, *Appl. Phys. Lett.*, **85**, 202-204 (2004).
7. Q. H. Li, Q. Wan, Y. X. Liang, and T. H. Wang, *Appl. Phys. Lett.*, **84**, 4556-4558 (2004).

8. J. H. He, C. L. Hsin, J. Liu, L. J. Chen, and Z. L. Wang, Adv. Mater., **19**, 781-784 (2007).

9. W. K. Hong, J. I. Sohn, D. K. Hwang, S.S. Kwon,G. J, S. Song, S. M. Kim, H. J. Ko, S. J.Park, M. E. Welland, T. Lee, Nano Lett., **8**, 950-956 (2008).

10. J. Yang and S. Li, Y. Zhao, X. A. Zhao, *J. Phys. Chem. C*, **113**, 4804–4808 (2009).

11. T. Kamiya, K. Tajima, K. Nomura, H. Yanagi, and H. Hosono, *Phys. Status Solidi A*, **205**, 1929 (2008).

12. Z. J. Yang, L. H. Wan, Y. J. Yu, Y. D. Wei and J. Wang, *J. Appl. Phys.*, **108**, 033704 (2010).

13. Q. Han, B. Cao, L. P. Zhou, G. J. Zhang, and Z. H. Liu, *J. Phys. Chem. C*, **115**, 3447–3452 (2011).

14. N. Marzari and D. Vanderbilt, *Phys. Rev. B*, **56**, 12847 (1997); I. Souza, N. Marzari and D. Vanderbilt, *Phys. Rev. B*, **65**, 035109 (2001).

15. P. Giannozzi, S. Baroni, N. Bonini, M. Calandra, R. Car, C. Cavazzoni, D. Ceresoli, G. L. Chiarotti, M. Cococcioni, I. Dabo, A. Dal Corso, S. Fabris, G. Fratesi, S. de Gironcoli, R. Gebauer, U. Gerstmann, C. Gougoussis, A. Kokalj, M. Lazzeri, L. Martin-Samos, N. Marzari, F. Mauri, R. Mazzarello, S. Paolini, A. Pasquarello, L. Paulatto, C. Sbraccia, S. Scandolo, G. Sclauzero, A. P. Seitsonen, A. Smogunov, P. Umari, R. M. Wentzcovitch, *J. Phys. Condens. Matter*, **21**, 395502 (2009).

16. D. Vanderbilt, *Phys. Rev. B*, **41**, 7892 (1990).

17. R. Landauer, *Phil. Mag.*, **21**, 853 (1970).

SITE DEPENDENT TRANSPORT PROPERTIES OF N-DOPED GRAPHENE NANORIBBONS WITH ZIGZAG EDGES

YANG HU, YOUSONG GU[†], XU SUN, XUEQIANG WANG

School of Materials Science and Engineering, University of Science and Technology Beijing, 100083, Beijing, P R China

The electronic structures and the electronic transport properties for nitrogen doped (zig-zag grapheme nanoribbons) z-GNRs were investigated by first-principles calculations. For the perfect Z-5-8 GNR, the band structure, DOS and transimission spectrum is almost symmetry around the Fermi level, and the I-V curve is almost linear. For the N-doped Z-5-8 GNRs, there exist gaps in the band structures and DOS plots, and conductance is lower than that of the perfect one. As the doping site moves from inner to edge, electrical conductance is decreased. Nitrogen doping at the edge has the greatest impact on the transport properties of Z-5-8 GNR. Therefore, it can be concluded that influences of nitrogen doping on the transport properties for z-GNRs is sites dependent.

1. Introduction

Graphene is a sheet of carbon atoms bound together in a network of repeating hexagons within a single plane [1]. Its intriguing electric and transport properties have attracted wide attention since its discovery in 2004 [2-6]. Graphene nanoribbons (GNRs) are nanometer-sized graphene, which have attracted a good deal of attention for recent years [7]. Due to the quantum confinement effect, GNRs are expected to have electronic prosperities similar to those of CNTs that can be unwrapped into GNRs [8]. Single-wall CNTs exhibit either metallic or semiconducting behavior depending on their chirality and diameter [9]. The electronic properties for GNRs are determined by the width and orientation, such as armchair (A) or zigzag (Z) [10]. It is widely known that: (a) All z-GNRs remains gapless. (b) a-GNRs are most strongly affected by width. (c) Edge effects are negligible for wide nanoribbons [11]. For CNR with zigzag edges, the existence of a very large density of states at the Fermi level is attributed to the non-bonding localized edge states, which was predicted with tight-binding (TB) model [12] and first-principles calculations [13,14]. Scanning tunneling

---

[†] Corresponding author, e-mail: yousongu@mater.ustb.edu.cn

microscopy (STM) images of graphene sheets revealed bright stripes along its edges, suggesting a high density of edge states near the Fermi level [15].

Nitrogen doping can change the electronic structures and transport properties of carbon materials, so they are typical substitution dopants in GNRs. The electronic properties of N-doped z-GNRs GNRs are affected strongly [16-18]. Experimentally, nitrogen doped graphene has been successfully realized by chemical vapor deposition [19] and plasma etching method [23], and demonstrated its applications in bioelectrochemistry and as biosensors.

In this work, we perform detailed study on how the isolated substitution N dopant influences the electronic and transport properties and demonstrate the current-voltage (I-V) characteristics of N-doped GNRs from first-principles calculations. We show that various substitutional sites can inject different impurity levels and get a variety of I-V curves.

2. Calculation Methods

Eletronic structure and transport properties were accomplished using the siesta/transiesta code [21] based on the DFT and non-equilibrium Green's function. All calculations were performed with a single-zeta plus polarization orbitals (SZP) basis set. For the exchange and correlation term, the generalized gradient approximation (GGA) was used as proposed by Perdew-Burke-Ernzerhof (PBE). Plane cutoff energy is chosen as 250 Ry, and self-consistent-field tolerance was 1.0×10^{-4}.

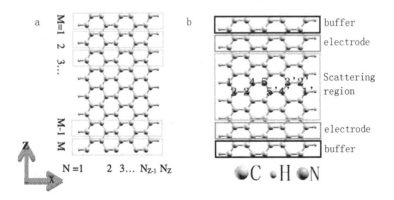

Figure 1. Atomic model of (a) z-GNR (b) N-doped z-GNR at different sites. The golden, blue and gray balls denote carbon, hydrogen and nitrogen atoms, respectively.

Supercells were adopted to mimic GNRs so that each ribbon is separated by vacuum regions. The intervals between the ribbons were kept at about 12 Å and 20 Å for edge-edge and layer-layer distances, respectively. These intervals were enough to ensure negligible interactions between the GNR and its periodic images. The supercell with one period of z-GNR had been relaxation firstly. Then supercells with multiple periods were constructed from it, in the sequence of the left buffer, left electrode, scattering region, right electrode and right buffer, as shown in Figure 1. The z-GNRs are named by the convention that the number of the zigzag chains (N) across the width and repeat units along the length (M). We refer to a z-GNR with N zigzag chains and M repeated units as a Z-N-M GNR.

In order to study the effect of N-doping, one carbon atom in the center was substituted by a nitrogen atom, and ten doping sites were studied as shown in Figure 1b. The positions of electrodes were constrained, while the positions of atoms in scattering region including the doping nitrogen atoms and passivated hydrogen atoms were fully relaxed under the condition that the cell parameters were fixed.

The band structure and density of states were calculated by the siesta program as a closed system and the transport properties were calculated by the transiesta program as an open system and post treated by the tbtran utility.

3. Results and Discussion

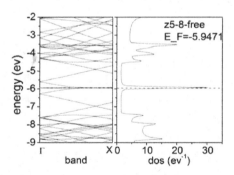

Figure 2. The electronic structures of Z-5-8 GNR: band structure (left) and corresponding total density of states (right). The dash line represents the Fermi level.

First, the electronic structure of hydrogenated undoped Z-5-8 GNR was calculated and the band structure and the density of states (DOS) were plotted in Figure 2. From Figure 2, we can see that the Z-5-8 GNR is metallic, which is similar to armchair carbon nanotube. However, there is a remarkably sharp peak which is bisected by the Fermi level. The sharp peak comes from the edge states contributed by unpaired π edge electrons [22], which results from a quantum confinement effect depending on the ribbon width [23].

There are 10 sites across the width from left edge to center and from the center to the right edge, as indicated in Figure 1(b). For comparison, the total energies of N-doped GNRs are plotted in Figure 3. As we can see, substitutional doping sites 1 and 1' (edge sites) are energetically favorable. Sites 2, 2', 3 and 3' (middle sites) are comparatively preferable. While 4, 4', 5 and 5' (inner sites) are the most unfavorable. The total energy increases as doping site moves from edge to inner.

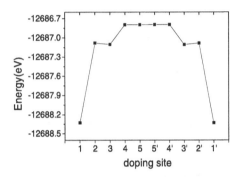

Figure 3. The total energies of N-doped GNR at different doping sites

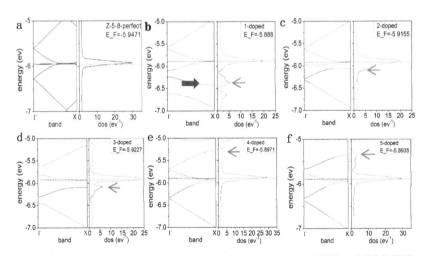

Figure 4. The band structure and density of states for the hydrogenated N doped Z-5-8 GNRs at different doping sites.

The band structures and density of states for the hydrogenated N-doped Z-5-8 GNRs at different doping sites are shown in Figure 4. The electronic structure for perfect Z-5-8 GNR is also shown in Figure 4(a) for comparison. From symmetry consideration, doping sites n and n' (n=1-5) were the same so that the band structure and density of states are identical. In the case of perfect z-GNR, there are two wide and smooth bands near the Fermi level and there are also two very narrow bands located at the Fermi level, which are correspond to the sharp peak in DOSs. Most of the sharp peak is above the Fermi level, which is not mentioned by the previous work. It is worth noting that there are bands appearing near the Fermi level, which also leads to sharp peaks in DOSs. Different from perfect Z-5-8 GNR, symmetries of electronic structures of all the nitrogen doped GNRs have been broken. For the GNR doped at edge site 1, the narrow bands degenerate into one band and located right at the Fermi level. For the GNR doped at middle site 2, the top of the wide band below the Fermi level move downward and there is a gap of about 0.08eV below the Fermi level. For the GNR doped at middle site 3, the wide band above the Fermi level becomes a narrow band with 0.15 eV, and the wide band below the Fermi level also narrowed. There are three gaps existed near the Fermi level. For the GNR doped at inner site 4, there is only a very narrow gap just below the Fermi level, two small gaps nearby and a large gap 1.0 eV above the Fermi level. For the GNR doped at inner site 5, the wide band just above the Fermi level narrowed and the wide band just above the Fermi level slightly narrowed. The two small gaps near the Fermi level and the large gap above the Fermi level become much wide as compared with the case of inner site 4. Therefore, it can be seen that nitrogen doping and its sites affect the electronic structures of z-GNRs significantly.

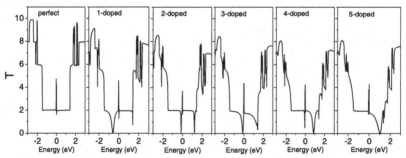

Figure 5. Transmission spectra at equilibrium of the hydrogenated perfect and nitrogen-doped Z-5-8 GNRs at different doping sites

The transmission spectra at equilibrium (V_{bias}=0V) of the perfect and nitrogen doped Z-5-8 graphene nanoribbons were calculated by the transiesta program and shown in Figure 5. It can be seen that the perfect Z-5-8 GNR's transmission coefficient is integer. There is a sharp peak at the Fermi level, and the lowest value is 2 in the vicinity of it. The sharp peaks near the zero energy correspond to the narrow bands near the Fermi level. When doped with nitrogen at different sites, the transmission coefficient dropt to almost 0 in some energy ranges, which corresponding the gaps in the band structures. In the entire energy range of all cases, transport coefficients of the N-doped GNRs are small than that of the perfect one, which is caused by the scattering of electrons by the doped N atoms.

I-V curves were calculated for the perfect and N-doped GNRs at different sites as shown in Figure 6. In the low bias voltages, the I-V curves are almost linear, as shown in the enlarged inset of Figure 6. A linear relationship is found for the perfect z-GNR, indicating a good metallic conductance. For N-doped Z-5-8 GNRs at site 4 and 5, I-V curves are approximately linear, but the currents is a little bit lower than that of the perfect one which indicates that nitrogen doping could hinder the transport of electrons. For N-doped Z-5-8 GNRs at site 1 and 2, I-V curves are approximately linear, but the currents dropt by a large amount as compared to the perfect one which indicates that nitrogen doping at the edge could block the transport of electrons. In the case of N-doping at site 3, the I-V curve is somewhat in the middle. Comparing the I-V characteristic curves of the perfect GNR and N-doped GNRs with dopant from center to edge, I-V curves' performances are gradually degraded. For example, in 2V, the current of perfect Z-5-8 GNR is 1.49×10^{-4} A, and the current of nitrogen doped Z-5-8 GNRs are 1.24, 1.28, 1.11, 0.85 and 0.72×10^{-4} A for doping at site from 5 to 1, respectively. The minimum current is about half of the perfect one. The large effect on nitrogen doping at edge is due to the fact that transport of electrons in GNR is mainly through the edge.

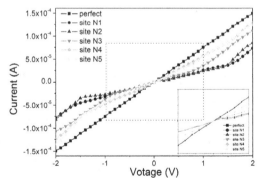

Figure 6. 2 I-V characteristic curves of the hydrogenated perfect and nitrogen-doped Z-5-8 GNRs at different doping sites.

4. Summary

The electronic structures and the transport properties of nitrogen doped Z-5-8 GNRs at different sites were obtained by calculation with the siesta/transiesta program. For the perfect Z-5-8 GNR, the band structure, DOS and transmission spectra is almost symmetry around the Fermi level, and the I-V curve is almost linear. For the N-doped Z-5-8 GNRs, there exist gaps in the band structures and DOS plots, and conductance is lower than that of the perfect one. As the doping site moves from inner to edge, the electrical conductance decreases. Nitrogen doping at the edge has the greatest impact on the transport properties of Z-5-8 GNR. Therefore, it can be concluded that influences of nitrogen doping on the transport properties for z-GNRs is sites dependent.

Acknowledgement

This work was supported by the Major Project of International Cooperation and Exchanges (2006DFB51000), NSFC (50972009), the Research Fund of Co-construction Program from Beijing Municipal Commission of Education, the Fundamental Research Funds for the Central Universities.

References

1. A.K. Geim, P. Kim, *Scientific American.* **4,** 90 (2008)
2. C. Berger, Z.M. Song, X.B. Li, X.S. Wu, N. Brown, C. Naud, D. Mayou, T.B. Li, J. Hass, A.N. Marchenkov, E.H. Conrad, P.N. First, W.A. De Heer, *Science* **312,** 1191 (2006).
3. M.Y. Han, B. Öyilmaz, Y.B. Zhang, P. Kim, *Phys. Rev. Lett.* **98,** 206805 (2007).
4. A. Rycerz, J. Tworzydlo, C.W.J. Beenakker, *Nat. Phys.* **3,** 172 (2007).
5. B. Huard, J.A. Sulpizio, N. Stander, K. Todd, B. Yang, D. Goldhaber-Gordon, *Phys. Rev. Lett.* **98,** 236803 (2007).
6. K.S. Novoselov, A.K. Geim, S.V. Morozov, D. Jiang, Y. Zhang, S.V. Dubonos, I.V. Grigorieva, A.A. Firsov, *Science* **306,** 666 (2004).
7. K. Wakabayashi, M. Fujita, H. Ajiki, and M. Sigrist, *Phys. Rev. B* **59,** 8271-8282 (1999).
8. V. Barone, O. Hod, and G. E. Scusera, *Nano Lett,* **6,** 2748-2754 (2006).
9. Hamada, N.; Sawada, S.; Oshiyama, A. *Phys. ReV. Lett.* **68,** 1579 (1992).
10. Y.-W. Son, M. L. Cohen, and S. G. Louie, *Phys. Rev. Lett.,* **97,** 216803-1–216803-4 (2006).
11. M. Ezawa, *Phys. Rev. B* **73,** 045432-1–045432-8 (2006).

12. Wakabayashi K, Fujita M, Ajiki H, Sigrist M. *Phys Rev B* **59**, 8271–82 (1999).
13. Miyamoto Y, Nakada K, Fujita M. *Phys Rev B* **59**, 9858–61 (1999).
14. Kawai T, Miyamoto Y, Sugino O, Koga Y. *Phys Rev B* **62**, 16349–52 (2000).
15. Y. Kobayashi, K.-I. Fukui, T. Enoki, and K. Kusakabe, *Phys. Rev. B* **73**, 125415 (2006).
16. T. B. Martins, R. H. Miwa, A. J. R. da Silva, and A. Fazzio, *Phys.Rev. Lett.*, **98**, 196803 (2007).
17. B. Huang, Q. M. Yan, G. Zhou, J. Wu, B. L. Gu, and W. H. Duan, *Appl. Phys. Lett.* **91**, 253122 (2007).
18. Shan Sheng Yu; Wei Tao Zheng; Qing Jiang, Nanotechnology, *IEEE Transactions on*, **9**, 78-81 (2010).
19. Wei D C; Liu Y Q; Wang Y; Zhang H L; Huang L P; Yu G, *Nano Lett*, **9**, 1752 (2009).
20. Ying Wang; Yuyan Shao; Dean W. Matson; Jinghong Li; and Yuehe Lin; *ACS Nano*, **4**, 1790-1798 (2010).
21. Brandbyge M; Mozos J L; Ordejon P; Taylor J; Stokbro K, 2002 *Phys. Rev. B* **65**, 165401
22. A. J. Read, R. J. Needs, K. J. Nash, L. T. Canham, P. D. Calcott, and A. Qteish, *Phys. Rev. Lett.* **70**, 2050 (1993).
23. Nakada K, Fujita M, Dresselhaus G, Dresselhaus MS. *Phys Rev B* **54**, 17954–60 (1996).

LENGTH EFFECT ON THE ELECTRONIC TRANSPORT PROPERTIES OF MG/ZNO NW/MG NANOSTRUCTURES STUDIED BY FIRST PRINCIPLES CALCULATION

YOUSONG GU[‡], REN GAO, XU SUN AND XUEQIANG WANG

Department of Material Physics and Chemistry, University of Science and Technology Beijing, Beijing 100083, People's Republic of China

First principles calculations were performed to study the electronic structures and transport properties of Mg/ZnO NW/Mg nanostructures and the length effect via the Siesta/Transiesta codes. It is found that when the length of the ZnO nanowires between the Mg electrodes is short, local density of states (LDOS) near the Fermi level distribute over the whole scattering region and good conductance, Ohmic contact were observed, due to charge redistribution at the interface of Mg and ZnO. When the length of ZnO nanowires becomes large, the LDOS near the Fermi level vanishes in the center of the ZnO nanowires, which result in rectifying I ~ V curves and Schottky contact. The length of the ZnO nanowire between the Mg electrodes plays an important role in the I ~ V characteristics of the Mg/ZnO/Mg nanostructures.

1. Introduction

Zinc oxide is a direct and wide band gap (E_g=3.37eV) semiconductor with high exciton binding energy (E_b=60meV). Due to its interesting electronic, piezoelectric and photoelectric properties, ZnO nanowires have a wide range of applications in electronic, piezo-electronic, optoelectronic devices and sensors [1-5]. The electronic transport properties of ZnO nanowires are extremely important in the design and optimization of nanodevices [6].

Many experimental results are available on the electronic transport properties of ZnO NWs [7]. Electronic transport properties of ZnO nanowires with doping and under strains are also reported [8, 9]. However, only a few theoretical works on the transport properties of ZnO nanowires were reported [10-12]. Kamiya et al. [10] calculated the electronic transport of bulk ZnO coupled by Au or Mg electrodes and found Schottky and Ohmic contacts at the Au/ZnO/Au and Mg/ZnO/Mg interfaces, respectively. Yang et al. [11] studied electronic transport of ZnO nanowires coupled by aluminum electrodes and

---

[‡] Corresponding author, e-mail: yousongu@mater.ustb.edu.cn

observed clearly rectifying current-voltage characteristics. The length dependence of electronic transport properties of single-walled ZnO nanotubes was investigated by Qin Han et al. [12] and it is found that the conductance decreases exponentially with the length of the nanotubes at low bias but the current are insensitive to the lengths at high bias.

In this work, first principles calculations based on density functional methods and non-equilibrium Green function (NEGF) were performed to study length dependence of the electric structure and transport properties of Mg/ZnO/Mg nanostructures.

2. Computational Detail

First-principles calculations were performed in the frame work of density functional theory using the Siesta/Transiesta package [13-14], which employed pseudopotentials and numerical atomic orbit basis sets. In the calculation, generalized gradient approximations in the form proposed by Perdew, Burke, and M. Ernzerhof (GGA-PBE) were chosen as the exchange correlation functionals [15], and the double zeta, polarized (DZP) basis sets were chosen. The mesh cutoff was 250 Ry and 1x1x30 grid were used for k-point sampling for electronic structure calculation. The supercell for nanowire was chosen as a big box with one unit of nanowire located in the center surround with plenty of spacing so that the c-axis length is the period of the nanowire and the separation between the nanowires is larger that 10Å. Geometry relaxation were performed before electronic structure calculation. A series of nanowires L1 ~ L6 corresponds to nanowires with one to six periods in the c-axis.

Figure 1. The atomic model used to calculate the transport properties of Mg/ZnO /Mg nanostructure.

In order to calculate the electronic transport properties of ZnO nanowires, Mg electrodes were assembled to the ZnO nanowires so that close contact and minimum total energy are realized by a series of geometry relaxations. The assembled system used for transport calculation was shown in Figure 1.

Transiesta calculation was performed for the electrode first, and then the whole system was calculated to get the Hamiltonian and overlaping (TSHS file) for each bias voltage. Transmission spectra and I ~ V curves were obtained by the tbtran post processing tool.

3. Results and Discussion

3.1. *Electronic structures*

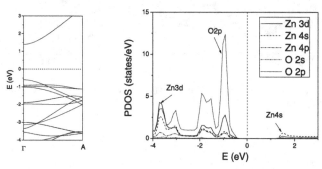

Figure 2. The band structure and projected density of states (PDOS) of ZnO nanowires. The dotted lines are the Fermi level.

The band structure and density of states were obtained by the Siesta program and shown in Figure 2. It is a typical semiconductor with direct and wide band gap. The Fermi levels are located at the middle of the gap and the calculated band gap is about 2.0eV which is small that experimental ones. The valence band is mainly composed of O 2p states, together with a small portion of hybrid Zn 3d states, and the conduction band is mainly composed of Zn 4s states.

The electronic structures of Mg/ZnO/Mg nanostructures were calculated after geometry relaxation by the SIESTA code. Figure 3 shows the atomic positions after geometry relaxation and the local density of states (LDOS) at Fermi level as shown by the iso-surface (isovalue=0.01 states/Bohr^3). It is clear that LDOS near the Fermi level is distributed all over the Mg electrodes and the interface layer of ZnO in the supercell due charge transfer. The LDOS near the Fermi level decreases rapidly with the distance from the interface. The LDOS near Fermi level in the center decreases quickly as the length of the nanowires increases.

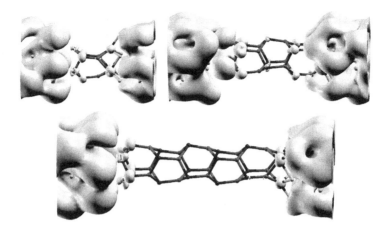

Figure 3. The local density of states (LDOS) at Fermi Level for Mg/ZnO/Mg with different nanowires lengths (a) L1 (b) L2 and (c) L3

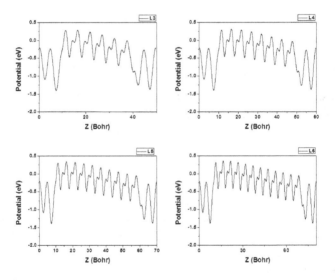

Figure 4. The I ~ V curves of ZnO NWs with different length (a) L3 (b) L4 (c) L5 and (d) L6.

The total potentials in the Mg/ZnO/Mg nanostructures are evaluated for models L3-L6 and averaged value in a plane are shown in Figure 4. It is not symmetric due to the polarization of ZnO along the c-axis. Potential drop is observed at the two ends of the nanowires due to charge redistribution at the interfaces. Polarized electric fields can be evaluated by the slopes of the potential envalope and the results are listed in Table 1. It can be found that the polarized electric field decreases as the nanowire length increases.

Table 1 The electronic field in Mg/ZnO/Mg nanostructures with different nanowire lengths.

Model	L3	L4	L5	L6
Zn site	0.0144	0.0116	0.0100	0.0086
O site	0.0145	0.0115	0.0099	0.0083
average	0.0145	0.0115	0.0099	0.0085

3.2. *Transport properties*

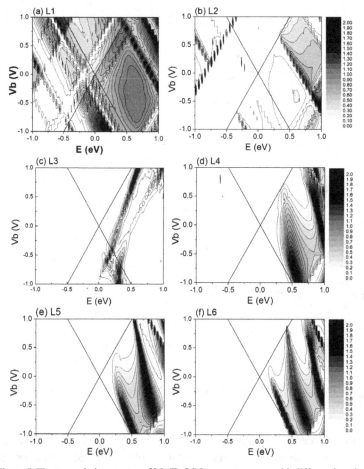

Figure 5. The transmission spectra of Mg/ZnO/Mg nanostructures with different lengths.

The electronic transport properties of ZnO nanowires were calculated by the Transiesta code. The transmission spectra are shown in Figure 5 and clear length dependence can be found. In transiesta calculation, electrons with energy in the range of -0.5e Vb ~ 0.5e Vb can be transmitted. In the case of L1, electrons can be transmitted easy except in some white valleys. As the length increases, more and more regions become white and it become hard for electrons to transport.

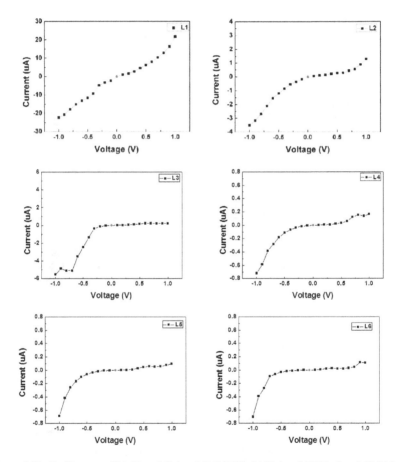

Figure 6. The I ~ V curves of Li, Na and K doped ZnO NWs. (a) Li doped (b) Na doped (d) K doped ZnO nanowires.

The calculated I ~ V characteristic curves are shown in Figure 6. It can be seen that almost linear I ~ V relationship is observed and Ohmic contact is assumed for model L1 since electrons are easy to transport through the short nanowire. As the length increase to L3 ~ L6, typical rectify I ~ V curves were found and Schottky contact is apparent. The cases of L2 is somewhat in between and can be regard as a poor conductor with Ohmic contact. The main parameters of the I~V curves are summarized in Table 2. It can be see that the conductance decreases from 19.54 to 3.66 as the number of periods increase from 1 to 2, and it turned into Schottky contact as the length increases further. The turn on voltage increase from L3 to L6 as the number of period increase from 3 to 6.

Table 2 The electronic transport characteristics for ZnO nanowires with different length between Mg electrodes

Models	L1	L2	L3	L4	L5	L6
contact	Ohmic	Ohmic	Schottky	Schootky	Schottky	Schottky
Conductance(Ω)	19.54	3.65	--	--	--	--
V_{turn_on} (V)	--	--	-0.3	-0.4	-0.5	-0.6

4. Conclusions

The electronic structures and transport properties of Mg/ZnO NW/Mg nanostructures and the length effect were studied via the Siesta/Transiesta codes. It is found that when the length of the ZnO nanowires between the Mg electrodes is short, local density of states (LDOS) near the Fermi level distribute over the whole scattering region and good conductance, Ohmic contact was observed, due to charge redistribution at the interface of Mg and ZnO. When the length of ZnO nanowires becomes large, the LDOS near the Fermi level vanishes in the center of the ZnO nanowires, which result in rectifying I ~ V curves and Schottky contact. The length of the ZnO nanowire between the Mg electrodes plays an important role in the I ~ V characteristics of the Mg/ZnO/Mg nanostructures.

Acknowledgement

This work was supported by the Major Project of International Cooperation and Exchanges (2006DFB51000), NSFC(50972009, 51172022), NSAF(10876001), the Research Fund of Co-construction Program from Beijing Municipal Commission of Education, the Fundamental Research Funds for the Central Universities.

References

1. Y. Yang, J. J. Qi, W. Guo, Y. S. Gu, Y. H. Huang and Y. Zhang, *Phys. Chem. Chem. Phys.*, **12**, 12415–12419 (2010).

2. X. M. Zhang, M. Y. Lu, Y. Zhang, L. J. Chen, and Z. L. Wang, *Adv. Mater.*, **21**, 2767–2770 (2009).

3. Y. Yang, J. J. Qi, Q. L. Liao, H. F. Li, Y. S. Wang, L. D. Tang and Y. Zhang, *Nanotechnology*, **20**, 125201 (2009).

4. Z. L. Wang and J. H. Song, *Science*, **312**, 242 (2006).

5. Y. Lei, X. Q. Yana, N. Luo, Y. Song, Y. Zhang, *Colloids and Surfaces A: Physicochem. Eng. Aspects*, **361**, 169–173 (2010).

6. Y. Hu, Y. Zhang, C. Xu, G. Zhu, and Z.L. Wang, Nano Lett. **10**, 5025(2010).

7. P.-X. Gao, Y. Ding and Z.L. Wang, Nano Lett. **9**, 137 (2009).

8. Y. Yang, J.J. Qi, Y. Zhang, Appl. Phys. Lett. **92**, 182117 (2008).

9. K.H. Liu, P. Gao, Z. Xu, Appl. Phys. Let. **92**, 213105 (2008).

10. T. Kamiya, K. Tajima, K. Nomura, H. Yanagi, and H. Hosono, *Phys. Status Solidi A*, **205**, 1929 (2008).

11. Z. J. Yang, L. H. Wan, Y. J. Yu, Y. D. Wei and J. Wang, *J. Appl. Phys,*. **108**, 033704 (2010).

12. Q. Han, B. Cao, L. P. Zhou, G. J. Zhang, and Z. H. Liu, *J. Phys. Chem. C*, **115**, 3447 (2011).

13. E. Artacho, E. Anglada, O. Dieguez, J. D. Gale, A. García, J. Junquera, R. M. Martin, P. Ordejón, J. M. Pruneda, D. Sánchez-Portal and J. M. Soler, J. Phys.: Condens. Matter, **20**, 064208 (2008).

14. J.M. Soler, E. Artacho, J.D. Gale, A. García, J. Junquera, P. Ordejón, and D. Sánchez-Portal, J. Phys. Condens. Matter, **14**, 2745 (2002).

15. J.P. Perdew, K. Burke, and M. Ernzerhof, Phys. Rev. Lett. **77**, 3865 (1996).

ORDERED ZNO NANOROD ARRAYS FOR ULTRAVIOLET DETECTION

FANG YI, YUNHUA HUANG, YOUSONG GU[§]

State Key Laboratory for Advanced Metals and Materials, School of Materials Science and Engineering, University of Science and Technology Beijing, Beijing 100083, People's Republic of China

ZHENG ZHANG, QI ZHANG

Key Laboratory of New Energy Materials and Technologies, University of Science and Technology Beijing, Beijing 100083, People's Republic of China

Well-aligned ZnO nanorod arrays were grown on the Si substrate via a hydrothermal method. Excited by a 325 nm laser, the ZnO nanorods radiate a focused beam of ultraviolet light. We briefly discuss the mechanisms for emissions in the photoluminescence spectrum. A simple and low-cost method was used to fabricate the metal-semiconductor-metal photodetector based on ZnO nanorod arrays. The current-voltage curve of the device shows double Schottky diode characteristics in the dark, and it transforms to Ohmic under ultraviolet illumination. The photogenerated current under 365 nm ultraviolet illumination is almost 25 times higher than that under 254 nm ultraviolet illumination, which is due to the easier recombination of electron-hole pairs under 254-nm ultraviolet light. The results in this paper provide a possible way to facilitate the fabrication process of nanodevices and shed light on the detection for ultraviolet of different wavelengths.

1. Introduction

With a direct band gap of 3.37 eV and a large exciton binding energy of 60 meV at room temperature, ZnO is a promising candidate for optoelectronic devices such as ultraviolet (UV) nanolasers[1,2], light emitting diodes[3,4], and UV photodetectors[5,6]. One-dimensional ZnO nanostructures are especially appealing due to their large surface-to-volume ratio and short carrier transit time. However, the fabrication process of optoelectronic devices is still too intricate for large-scale manufacture such as the "pick and place" method, the photolithography technology, and the spin-coating step of photosensitive resist. Furthermore, numerous reports have studied the photoresponse of one-

---

[§] Corresponding author, e-mail: yousongu@mater.ustb.edu.cn

dimensional ZnO nanostructures to UV light, but the wavelength selectivity has not gained considerable attention yet.

In this paper, we fabricate the well-aligned ZnO nanorod arrays based metal-semiconductor-metal (MSM) photodetector through a simple and low-cost method, by which two Au electrodes are directly sputtered on opposite sides of the ZnO nanorod arrays. The I-V characteristics as well as the differences between photoresponse to shot-wavelength UV (254 nm) and photoresponse to near-band-edge wavelength UV (365 nm) were studied.

2. Experimental details

The ZnO nanorod arrays were synthesized on a p-Si (<100>, $\rho=15\sim30$ $\Omega\cdot$cm) substrate through a hydrothermal method. Before the hydrothermal reaction, the silicon wafer was cleaned by organic solvents and etched in dilute HF acid. Then, a ZnO seed layer was deposited on the substrate by a spin-coating technique and subsequently annealed at 350 °C for 30 minutes to evaporate the solvent and remove the organic residuals. During the hydrothermal process, the ZnO nanorods were grown in an aqueous solution of zinc nitrate hydrate (0.05 M) and hexamethylenetetramine (0.05 M) at 95 °C for 15 hours. After the reaction, the substrate with ZnO nanorods was removed from the solution, rinsed with distilled water, and dried at 70 °C for 3 hours. Finally, a 100-nm-thick Au film was sputtered on opposite sides of the ZnO nanorod arrays with a shadow mask. The structure of the fabricated MSM photodetector is depicted in Figure 1a.

The morphology of the ZnO nanorod arrays was observed by a field emission scanning electron microscope (FESEM). The crystal structure was analyzed by X-ray diffraction (XRD). The photoluminescence (PL) was examined using a He-Cd (325 nm) laser as the excitation source. Current-voltage (I-V) characteristics were measured by a semiconductor analyzer (Keithley 4200). UV-visible absorption spectrum was measured at room temperature with a UV-visible spectrophotometer. The light sources for the photoresponse were provided by a hand-held UV lamp.

3. Results and discussion

Figure 1b and 1c show the FESEM images of the ZnO nanorod arrays. It can be seen that well-aligned ZnO nanorod arrays were vertically grown on the Si substrate and very closely packed. Some of the nanorods were ruptured during increasing the template. The average length of the nanorods is estimated to be 2 μm and the diameter is in the range of 50-100 nm, with clear hexagonal structure.

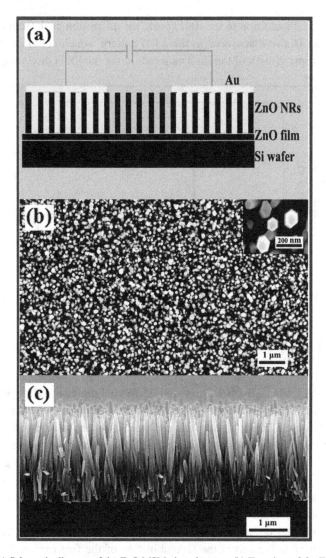

Figure 1. (a) Schematic diagram of the ZnO MSM photodetector. (b) Top view of the ZnO nanorod arrays over a large surface. The inset shows the hexagonal tip of the nanorods. (c) Cross-sectional FESEM image of the ZnO nanorod arrays.

Figure 2 shows the XRD pattern with 2θ from $20°$ to $60°$ of the ZnO nanorod arrays grown on the Si substrate. The observed diffraction peak can be indexed to that of the hexagonal wurtzite ZnO. Much stronger intensity and narrow spectral width of the (001) peak suggests the good crystallinity of the

ZnO nanorods. In addition to the (001) peak, the diffraction peak from the (100) crystal planes is also discovered in the XRD pattern, which could be attributed to the anisotropic growth of the ZnO nanorods along the [001] direction[7].

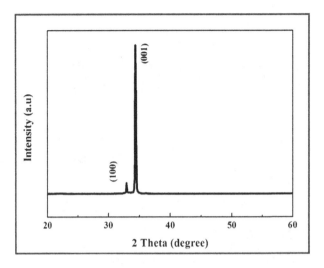

Figure 2. XRD pattern of the ZnO nanorod arrays. The prominent (001) peak demonstrates high crystallinity and dominating vertical alignment of the nanorods.

The solid line in Figure 3a presents the room-temperature PL spectrum of ZnO nanorod arrays coated on the Si substrate, which shows a strong UV emission at about 377 nm and a weak red emission at about 753 nm. The dash line in Figure 3a is the room-temperature PL spectrum of the pure Si wafer, a very weak broad emission around 450-650 nm is observed, as magnified in Figure 3c. The ZnO nanorods and the pure Si wafer were excited by a 325 nm laser with the same power. Figure 3b is the PL image of ZnO nanorods, which exhibits a focused beam of UV light.

The sharp and dominate UV emission corresponds to the near-band-edge emission and is derived from the recombination of the free excitons[8], which demonstrates that the ZnO nanorods have high crystal structure[9,10]. Considering the band gap of ZnO is 3.37 eV (or 368 nm), there is a red shift between the UV emission (377 nm or 3.29 eV) and the band gap, which is due to the Stokes shift[11].

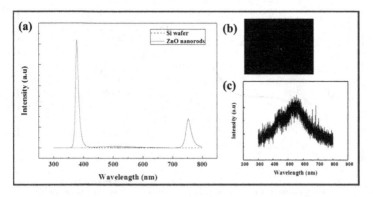

Figure 3. (a) PL spectra of the ZnO nanorods and the pure p-Si wafer. (b) PL image of the ZnO nanorods. (c) Magnified PL spectrum of the pure p-Si wafer.

Besides the UV emission, other emissions observed in the PL spectrum of ZnO nanorods are mostly related to transitions involving defect levels, indicating the existence of native defects. The native or intrinsic defects in ZnO mainly contain zinc or oxygen vacancies, zinc or oxygen interstitials, zinc or oxygen antisites. The oxygen vacancy is considered to have the lowest formation energy among the donors[12,13] and the oxygen antisite is regarded to have the highest formation energy among the acceptors[14]. The defect-related emissions are almost inevitable, especially the wide broad green luminescence. The mechanism for the green emission has been debated for a long time, and there is still no satisfactory consensus on this issue. There have been reports that attributed the broad green illumination to different defects, such as the recombination of photogenerated holes with the singly ionized oxygen vacancies[15-19], transitions between electrons in the conduction band and zinc vacancy levels[20-22], the electron transition from the bottom of the conduction band to the antisite defect O_{Zn} level[23], and zinc interstitials[24,25].

The synthesized ZnO nanorod arrays show almost no emission ranging from 420 nm (or 2.95 eV) to 720 nm (or 1.72 eV), which means the ZnO nanorods have hardly any defects of singly ionized oxygen vacancies (~ 2 eV[26]), zinc vacancies (2.5 eV[14]) and O_{Zn} antisites (2.38 eV[23]). The weak red emission around 753 nm may not result from the defects in ZnO nanorods[27-29] but correspond to the second-order feature of UV band-edge emission[30].

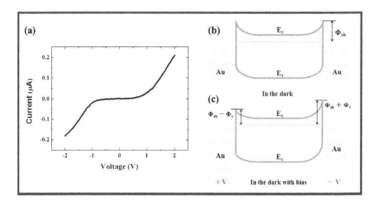

Figure 4. (a) I-V characteristics of the device biased at 2 V in the dark. Schematic energy band diagrams of the two Schottky contacts in the dark (b) with no bias and (c) with applied bias.

The work function of Au is 5.1 eV, which is much higher than the work function of ZnO, so theoretically Au will form Schottky contact with ZnO. As is shown in Figure 4a, the I-V characteristics of the fabricated ZnO MSM photodetector in the dark reveal double Schottky diode characteristics, whereas the two Au/ZnO junctions both form Schottky contacts in our experiment. Figure 4b and 4c are the energy band diagrams for the two Schottky barriers that are formed at the Au/ZnO interfaces with no bias and with applied bias, respectively.

The electron affinity of ZnO is 4.35 eV[31], so the theoretical Schottky barrier height between Au and ZnO could be 0.75 eV by using the expression, $\Phi_{B0} = (\Phi_m - \chi)$. However, the interface states have a great impact on the value of the barrier height and the published reports about the value of the Au Schottky barrier height on bulk ZnO crystals is close to 0.66 eV[32-34]. As for ZnO nanorods, the large surface-to-volume ratio and the high density of surface trap states make it more complicated to determine the accurate value of the Au Schottky barrier height, which we would not discuss in this report.

The symmetric I-V characteristics of the device in the dark confirm the uniformity of the ZnO nanorod arrays and the good contact between the two Au electrodes and ZnO nanorod arrays. At an applied bias of 2 V, the dark current is approximately 0.2 μA. The small dark current should be attributed to the highly resistive nature of the undoped ZnO nanorods and the large Au Schottky barrier height.

Figure 5a and 5b represents the saturated photocurrent of the fabricated device exposed to UV light with 2 V applied bias, the 254 nm UV and the 365 nm UV have the same power density of 1.60 mW/cm^2. The solid line in

Figure 5a is the I-V characteristics of the device upon 254 nm UV illumination, it can be seen that the saturated photocurrent is about 1.38 μA and the Schottky diode characteristics become less pronounced. The solid line in Figure 5b is the I-V characteristics of the device upon 365 nm UV illumination, it can be seen that the saturated photocurrent is about 30 μA and the I-V curve becomes linear, which means the Au/ZnO contacts change from Schottky to Ohmic.

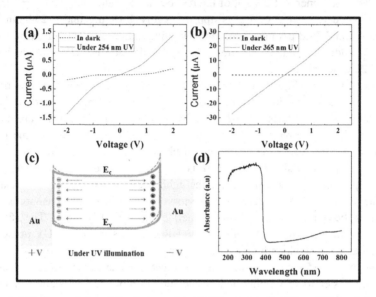

Figure 5. I-V characteristics of the device under (a) 254 nm and (b) 365 nm UV irradiation. (c) Schematic energy band diagram of the Schottky contacts under UV irradiation. (d) UV-visible absorption spectrum of ZnO nanorod arrays coated on the Si wafer.

For one-dimensional ZnO nanostructures based photodetectors, it is generally accepted that the photoconduction in ZnO is governed by absorption and desorption of oxygen molecules on the surface[5,35-40]. In the dark, the ZnO surface adsorbs oxygen molecules from the environment by providing free electrons $[O_2(g) + e^- \rightarrow O_2^-(ad)]$, therefore a depletion layer with low conductivity is formed near the surface. When exposed to UV light with photon energies higher than the band gap, electron-hole pairs are generated in ZnO [hυ $\rightarrow e^- + h^+$]. The holes migrate to the surface and discharge the absorbed oxygen ions [$h^+ + O_2^-(ad) \rightarrow O_2(g)$], causing desorption of the oxygen. The unpaired electrons flow towards the anode under an applied electric field and accumulate gradually until desorption and adsorption of O_2 achieve an equilibrium state, which leads to an increasing current until saturation.

Since the fabricated device consists of two back-to-back Schottky junctions, the barriers at the Au/ZnO interfaces obstruct the transport of the charge carriers and contribute to the small dark current. While the device is under UV illumination, the increased charge carriers trapped at the Au/ZnO interfaces would lead to a reduction of the Schottky barrier height. The energy band diagram for the contacts under UV illumination is depicted in Figure 5c.

The photogenerated current of the device can be calculated as $I_{pgc} = (I_{light} - I_{dark})$, hence the value of I_{pgc} is approximately 1.18 μA under 254 nm UV irradiation and 29.8 μA under 365 nm UV illumination, respectively. Compared with I_{pgc} under 254 nm UV illumination, I_{pgc} under 365 nm UV illumination is nearly 25 times higher.

As is shown in Figure 5d, there is a sharp absorption edge around 368 nm in the UV-visible absorption spectrum of the ZnO nanorod arrays coated on the Si substrate, and the absorbance values at 365 nm UV and 254 nm UV are nearly the same. The ZnO nanorod arrays have a very large surface-to-volume ratio, which brings about a higher absorption of photons than the bulk ZnO. However, because of quantum size effects, there is a substantial density of states near the band gap edges and an increasing recombination of excitons due to carrier confinement[1]. Thus the large surface-to-volume ratio of ZnO nanorods also enhances the recombination of photogenerated electron-hole pairs. The 365 nm UV has a wavelength near the absorption edge, while the 254 nm UV possesses a much shorter wavelength above the absorption edge, so the electron-hole pairs excited by the 254 nm UV have higher energies. The higher-energy electron pairs are easier to recombine and would induce more recombination of other electron-hole pairs near the surface, where has a huge density of states. What is more, the photons of 365 nm UV have a longer penetration depth than the photons of 254 nm UV, which means the photogenerated electron-hole pairs under 254 nm UV have a longer distance to travel across before they get to the electrodes, therefore more electron-hole pairs recombine during the longer travel. The larger percentage of recombined electron-hole pairs contributes to the much lower photocurrent when the device is under 254 nm UV illumination. The big difference in photoresponse to the two kinds of UV may make ZnO nanorods possible to selectively detect UV of different wavelengths.

4. Conclusion

In summary, the well-aligned ZnO nanorods showed a sharp and dominate UV emission, which indicates an extremely low level of native defects. A simple and low-cost way was adopted to fabricate the UV photodector. The I-V curve of the device showed symmetric Schottky diode characteristics in the dark and transformed to Ohmic under UV illumination. The photogenerated current under

365 nm UV illumination was much higher than that under 254 nm UV illumination, which is owing to the easier recombination of electron-hole pairs upon 254 nm UV irradiation. This report provides a possible way to facilitate the fabrication process of nanodevices and demonstrates the possibility of ZnO nanorods to selectively detect UV of different wavelengths.

Acknowledgments

This work was supported by the Major Project of International Cooperation and Exchanges (2006DFB51000), NSFC (51172022), NSAF (10876001), the Research Fund of Co-construction Program from Beijing Municipal Commission of Education and the Fundamental Research Funds for the Central Universities.

References

1. M. H. Huang, S. Mao, H. Feick, H. Yan, Y. Wu, H. Kind, E. Weber, R. Russo, P. Yang, *Science* **292**, 5523 (2001).
2. D. J. Gargas, M. C. Moore, A. Ni, S. W. Chang, Z. Zhang, S. L. Chuang, P. Yang, *ACS nano* **4**, 6 (2010).
3. X. M. Zhang, M. Y. Lu, Y. Zhang, L. J. Chen, Z. L. Wang, *Adv. Mater.* **21**, 27 (2009).
4. M. T. Chen, M. P. Lu, Y. J. Wu, J. Song, C. Y. Lee, M. Y. Lu, Y. C. Chang, L. J. Chou, Z. L. Wang, L. J. Chen, *Nano lett.* **10**, 11 (2010).
5. H. Kind, H. Yan, B. Messer, M. Law, P. Yang, *Adv. Mater.* **14**, 2 (2002).
6. S. M. Peng, Y. K. Su, L. W. Ji, C. Z. Wu, W. B. Cheng, W. C. Chao, *J. Phys. Chem. C* **114**, 7 (2010).
7. A. Dev, S. K. Panda, S. Kar, S. Chakrabarti, S. Chaudhuri, *J. Phys. Chem. B* **110**, 29 (2006).
8. Y. C. Kong, D. P. Yu, B. Zhang, W. Fang, S. Q. Feng, *Appl. Phys. Lett.* **78**, 4 (2001).
9. M. Foley, *Appl. Phys. Lett.* **93**, 24 (2008).
10. S. Lima, F. A. Sigoli, M. Jafelicci Jr, M. R. Davolos, *Internat. J. Inorg. Mater.* **3**, 7 (2001).
11. K. P. O'Donnell, R. W. Martin, P. G. Middleton, *Phys. rev. lett.* **82**, 1 (1999).
12. F. Oba, A. Togo, I. Tanaka, J. Paier, G. Kresse, *Phys. Rev. B* **77**, 24 (2008).
13. A. Janotti, C. G. Van de Walle, *J. crystal growth* **287**, 1 (2006).
14. A. Janotti, C. G. Van de Walle, *Phys. Rev. B* **76**, 16 (2007).
15. S. A. Studenikin, N. Golego, M. Cocivera, *J. Appl. Phys.* **84**, 4 (1998).
16. S. B. Zhang, S. H. Wei, A. Zunger, *Phys. Rev. B* **63**, 7 (2001).
17. K. Vanheusden, W. L. Warren, C. H. Seager, D. R. Tallant, J. A. Voigt, B. E. Gnade, *J. Appl. Phys.* **79**, 10 (1996).
18. F. Kröger, *J. Chem. Phys.* **22**, 2 (1954).

19. F. H. Leiter, H. R. Alves, A. Hofstaetter, D. M. Hofmann, B. K. Meyer, *phys. status solidi (B)* **226**, 1 (2001).
20. T. Sekiguchi, N. Ohashi, Y. Terada, *Jpn. J. Appl. Phys.* **36**, 3A (1997).
21. A. F. Kohan, G. Ceder, D. Morgan, C. G. Van de Walle, *Phys. Rev. B* **61**, 22 (2000).
22. D. C. Reynolds, D. C. Look, B. Jogai, J. E. Van Nostrand, R. Jones, J. Jenny, *Solid State Commun.* **106**, 10 (1998).
23. B. Lin, Z. Fu, Y. Jia, *Appl. Phys. Lett.* **79**, 7 (2001).
24. X. Liu, X. Wu, H. Cao, R. P. H. Chang, *J. Appl. Phys.* **95**, 6 (2004).
25. E. G. Bylander, *J. Appl. Phys.* **49**, 3 (1978).
26. A. Janotti, C. G. Van de Walle, *Appl. Phys. Lett.* **87**, (2005).
27. R. Cross, M. M. D. Souza, E. M. Sankara Narayanan, *Nanotechnology* **16**, (2005).
28. M. Gomi, N. Oohira, K. Ozaki, M. Koyano, *Jpn. J. Appl. Phys., Part 1*, 42 (2003).
29. R. B. Lauer, *J. Phys. Chem. Solid* **34**, 2 (1973).
30. T. Mahalingam, K. M. Lee, K. H. Park, S. Lee, Y. Ahn, J. Y. Park, K. H. Koh, *Nanotechnology* **18**, (2007).
31. J. A. Aranovich, D. Golmayo, A. L. Fahrenbruch, R. H. Bube, *J. Appl. Phys.* **51**, 8 (1980).
32. A. Y. Polyakov, N. B. Smirnov, E. A. Kozhukhova, V. I. Vdovin, K. Ip, Y. W. Heo, D. P. Norton, S. J. Pearton, *Appl. Phys. lett.* **83**, (2003).
33. B. J. Coppa, R. F. Davis, R. J. Nemanich, *Appl. Phys. lett.* **82**, (2003).
34. R. C. Neville, C. A. Mead, *J. Appl. Phys.* **41**, 9 (1970).
35. Q. H. Li, T. Gao, Y. G. Wang, T. H. Wang, *Appl. Phys. Lett.* **86**, (2005).
36. Q. H. Li, Q. Wan, Y. X. Liang, T. H. Wang, *Appl. Phys. Lett.* **84**, 22 (2004).
37. Q. H. Li, Y. X. Liang, Q. Wan, T. H. Wang, *Appl. Phys. Lett.* **85**, (2004).
38. Z. Fan, P. Chang, J. G. Lu, E. C. Walter, R. M. Penner, C. Lin, H. P. Lee, *Appl. Phys. Lett.* **85**, 25 (2004).
39. S. Kumar, V. Gupta, K. Sreenivas, *Nanotechnology* **16**, (2005).
40. C. Soci, A. Zhang, B. Xiang, S. A. Dayeh, D. Aplin, J. Park, X. Y. Bao, Y. H. Lo, D. Wang, *Nano lett.* **7**, 4 (2007).

EFFECT OF LOCALIZED UV IRRADIATION ON TRANSPORT PROPERTY IN ZNO NANOTETRAPOD DEVICES

WENHUA WANG, JUNJIE QI*, QINYU WANG, ZI QIN, ZENGZE WANG, XU SUN, FANG YI

Department of Materials Physics and Chemistry and State Key Laboratory for Advanced Metals and Materials, University of Science and Technology Beijing, 30 Xueyuan Road, Beijing 100083, China

Semiconductor optoelectronic devices based on a single ZnO nanotetrapod were constructed with Ohmic contact characteristics and the effect of localized UV irradiation on transport property in ZnO nanotetrapod device has been investigated. The measurements for the I-V characteristics and time-resolved measurements of current were conducted. The results indicate that the irradiation under UV light irradiation at the third leg of the tetrapod can readily tune the electrical transport property of the tetrapod along with favorable repeatability and reversibility. The current becomes larger as the UV light power density increases. A probable mechanism has been proposed and discussed. The ZnO nanotetrapod could be potentially used as detectors in irradiation environments.

1. Introduction

Currently, scholars are making massive researches on Zinc Oxide (ZnO) due to its wide band gap of 3.37 eV and large excitation energy of 60 meV at room temperature. Compared with other wide band gap semiconductors, ZnO has several advantages such as excellent electrical conductivity, unusual thermal conductivity, chemical stability and biological compatibility. A variety of ZnO crystal morphologies have been synthesized [1], such as nanowire, nanobelt, nanorod array, nanotetrapod [2]. ZnO nanotetrapod consists of four needle-shaped legs with the [0001] wurtzite structure bounded at the ZnO central core along the <111> of the zinc blende structure [3]. The angle between two adjacent legs is about 109°. And the preferential growth of the wurtzite phase along [0001] polar axis is the main reason for the formation of this tetrahedral structure together with an occupied 3-D tetrahedral space. Due to the specific structure, the tetrapod can bring an extensive range of applications [4]. So far, ZnO nanotetrapod has been widely used as field effect transistors [5], gas

---

* Corresponding author, e-mail: junjieqi@mater.ustb.edu.cn

118

sensors [6,7], UV detectors [7], p-n junction diodes [8], logic switches [9] and Schottky photodiodes [10]. However, there is no study on the response to UV light irradiation at the third leg of the tetrapod. Therefore, it is necessary to do some researches in this aspect.

In this paper, we constructed a semiconductor optoelectronic device based on a single ZnO nanotetrapod with Ohmic contact characteristic and evaluated the effect of UV light irradiation at the third leg of the tetrapod parallel to the substrate. The results indicate that the electrical transport property of the tetrapod can be tuned by the irradiation at the third leg of the tetrapod and the current response increases with the increasing of the UV light intensity. The possible mechanism for this phenomenon was also discussed.

2. Experimental

The ZnO nanotetrapods were synthesized by a vapor solid deposition process [11] and released from the substrate by ultrasonication treatment in alcohol. The single ZnO nanotetrapod-based semiconductor optoelectronic device with Ohmic contact characteristics was constructed as follows: ZnO nanotetrapod dispersion in alcohol was casted onto a silicon wafer covered with an insulating oxide layer of 500 nm in thickness, and then conducting silver glue was used to fix two adjacent legs among the three legs which lie approximately parallel to the substrate, as illustrated in Fig. 1(a). The UV light was introduced onto the other leg parallel to the substrate (the third leg of the tetrapod). Fig. 1(b) shows the Schematic diagram of the irradiation effect with UV light at the third leg of the tetrapod. Semiconductor characteristics system (Keithley 4200) which is equipped with pre-amplifier was employed to apply the voltage and measure the current of the system. A He-Cd laser (with a wavelength of 325 nm) was used as the source for the photo-excitation to test the UV irradiation response at atmosphere.

Fig. 1. (a) SEM image of the ZnO tetrapod in Ohmic contact with two Ag electrodes. (b) Schematic diagram of the system setup.

3. Results and discussion

The Current-Voltage (I-V) performance of the device based on a single ZnO tetrapod was firstly investigated and a He-Cd laser (with focused spot diameter less than 10 μm and wavelength of 325 nm) was used as the irradiation source to stimulate the third leg of ZnO tetrapod in the device at atmosphere. Linear I-V characteristic was obtained under dark when a voltage from -10V to 10V was applied, as shown in Fig. 2, indicating good Ohmic contact between Ag electrode and ZnO tetrapod. This is in complete accordance with the theoretical band structures, because the work function of Ag is 4.26 eV and the electron affinity of ZnO is 4.5 eV [12]. In order to investigate the effect of UV light irradiation on the electrical properties, the UV light was focused at the third leg of the tetrapod and then the I-V behavior in the same voltage range was measured. As illustrated in Fig. 2, the current increase upon the irradiation of UV light is seen clearly. According to the calculation, the resistance of the system decreases from 1.45 MΩ to 519 KΩ. To investigate reversibility of the irradiation response, a time-resolved measurement of the UV light irradiation response was conducted at a fixed voltage of 2 V. As can be seen in the inset at bottom right corner of Fig. 2, the device manifests fast recovery as well as good repeatability.

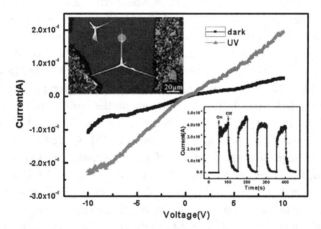

Fig. 2. *I-V* characteristics with and without UV light irradiation for the device in Ohmic contact with two Ag electrodes, the experimental setup at the top left inset. The bottom right corner inset shows the current response with intermittent UV irradiation.

Fig. 3 shows the time-resolved measurement of the UV light irradiation response at the same position of the third leg with the UV light irradiation power density

120

varying. Increasing the UV light power density leads to a significant rise in the current at a fixed voltage of 2V. It also can be seen that the irradiation response is reversible with a well repetition.

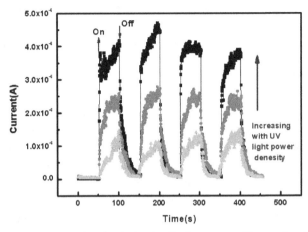

Fig. 3. The current response for the device in Ohmic contact with two Ag electrodes with intermittent UV illumination at the same position as UV illumination power density varies.

The above experimental results demonstrate that the effect of UV light irradiation at the third leg of the tetrapod could be detected in real time with ease by monitoring the current flow through the other two neighboring legs. A significant response can be yielded along with favorable repeatability and reversibility. As to the reason, a possible mechanism was brought forward and discussed. When the device was measured at atmosphere where is abundant with oxygen under dark, oxygen molecule adsorb on the surface by capturing free electrons from the n-type ZnO nanotetrapod, thus the surface depletion layer with low conductivity is formed near the surface [13]. While the UV light with photo energy larger than the band gap of ZnO is introduced to the third leg of the tetrapod, the electron-hole pairs are generated at the local regions [14, 15]. The photo-generated holes move to the surface of ZnO nanotetrapod and then be trapped by the absorbed oxide ions causing desorption of oxygen molecule, leading to an enhancement of the conductance [16]. The electrons migrate along the leg to the crystal nucleus and then subsequently to other legs of ZnO tetrapod under the bias of the outer circuit, giving rise to an increase of carrier density. Both the enhancement of conductance and the increase of the carrier density are responsible for the large increase in current upon localized UV light irradiation. Increasing laser power density affords direct increase of the light intensity, and the photo-generated carriers grow in number as the increase of the UV light intensity, therefore, the current response is enhanced.

4. Conclusion

In conclusion, we have investigated the effect of UV light irradiation on the third leg of the tetrapod for the Ohmic contact-typed device based on a single ZnO nanotetrapod. The measurement for the I-V characteristics and the time-resolved irradiation response was conducted. The conclusion is drawn that the current response can be tuned by UV irradiation at the third leg of the tetrapod along with favorable repeatability and reversibility and the current can also be affected by the UV light power density. This holds great potential for detections in UV light irradiation environments.

Acknowledgement

This work was supported by the National Basic Research Program of China (2007CB936201), the Major Project of International Cooperation and Exchanges (2006DFB51000), the National Natural Science Foundation of China (No. 50872008), the Research Fund of Co-construction Program from Beijing Municipal Commission of Education and the Fundamental Research Funds for the Central Universities.

References

1. Z. L. Wang, J. Phys.: *Condens. Matter* **16**, R829, (2004).
2. Y. Dai, Y. Zhang, Q.K. Li and C.W. Nan, *Chem. Phys. Let.* **358**, 83, (2002).
3. S.Takeuchi, H. Iwanaga and M. Fujii, *Philos. Mag.* **69**, 1125. (1994).
4. M. C. Newton and P. A. Warburton, *Materials Today* **10**, 50, (2007).
5. Y. D. Gu, J. Zhou, W. J. Mai, Y. Dai, G. Bao and Z. L. Wang, *Chem. Phys. Lett.* **484**, 96, (2010).
6. Z. X. Zhang, L. F. Sun, Y. C. Zhao, Z. Liu, D. F. Liu, L. Cao, B. S. Zou, W. Y. Zhou, C. Z. Gu and S. S. Xie, *Nano Lett.* **8**, 652, (2008).
7. L. Chow, O. Lupan and G. Y. Chai, *Phys. Status Solidi B.* **7**, 247, (2010).
8. M. C.Newton and R.Shaikhaidarov, *Appl. Phys. Lett.* **94**, 153112, (2009).
9. K. Sun, J. J. Qi, Q. Zhang, Y. Yang and Y. Zhang, *Nanoscale* **3**, 2166, (2011).
10. M. C. Newton, S. Firth and P. A. Warburton, *Appl. Phys. Let.* **89**, 072104, (2006).

11. H. F. Li, Y. H. Huang, Y. Zhang, J. J. Qi, X. Q. Yan, Q. Zhang, and J. Wang, *Crystal Growth & Design*, **9**, 4, (2009).
12. Z. L. Wang and J. H. Song, *Science*, **312**, 242, (2006).
13. J. Zhou, Y. D. Gu, Y. F. Hu, W. J. Mai, P. Yeh, G. Bao, A. Sood, D. L. Polla and Z. L. Wang, *Appl. Phys. Lett.* **94**, 191103, (2009).
14. T. Kamiya, K. Tajima, K. Nomura, H. Yanagi and H. Hosono, *Phys. Stat. Sol.* **205**, 1929, (2008).
15. A. Bera and D.Basak, *Appl. Phys. Lett.* **93**, 053102, (2008).
16. C. Soci, A. Zhang, B. Xiang, S. A. Dayeh, D. P. R. Aplin, J. Park, X. Y. Bao, Y. H. Lo, and D. Wang, *Nano Lett.* **7**, 1003, (2007).

ZNO NANOWIRES BASED MSM ULTRAVIOLET PHOTODETECTORS WITH PT CONTACT ELECTRODES

HANSHUO LIU, XIAOQIN YAN*, SIWEI MA, ZHIMING BAI

Department of Materials Physics and Chemistry, School of Materials Science and Engineering, University of Science and Technology Beijing, 30 Xueyuan Road, Beijing 100083, China

ZnO nanowires Metal-Semiconductor–Metal (MSM) photodetector with platinum (Pt) contact electrodes was fabricated and its optoelectronic properties were examined. The ZnO nanowires used in the experiment were synthesized by chemical vapor deposition method. The electrical performance and photoelectric response performance were studied, and the results showed that the Pt/ZnO nanowires based MSM ultraviolet photodetector exhibited a high sensitivity to 365 nm ultraviolet light, the photocurrent to dark current ratio can reach 9.44×10^3 with a voltage bias of 5 V, and the response and recovery time were also fast. Schottky barrier mechanism was employed to explain the results.

1. Introduction

Ultraviolet (UV) photodetector is becoming more and more important in numorous areas, such as UV astronomy, UV radiation dosimetry, flame detection, and water purification [1]. Recently, there has been increasing interest in zinc oxide, an attractive wide direct band gap (3.37 eV at room temperature) oxide semiconductor with a large excitation binding energy of 60 meV, which is suitable for fabrication of UV photodetectors [2, 3]. There have been researches on ZnO Schottky photodetectors and metal-semiconductor-metal (MSM) photodetectors in the UV region [4]. Platinum (Pt) has been used as a stable Schottky contact to ZnO due to its high metal work function (5.65 eV). Several researches on deposition ZnO Schottky diodes with Pt contact electrodes have been reported using silion, glass, plastic or other substrate [5, 6]. Meanwhile, compared with their bulk or thin film counterparts, one-dimensional ZnO nanostructures are expected to exhibit high photosensitivity due to their high surface-to-volume ratios and carrier confinement in two dimensions. A number

---

*Corresponding author's E-mail: xqyan@mater.ustb.edu.cn

of investigations on metal-semiconductor-metal photodetectors based on one-dimension based ZnO have been reported [7~9].

In this letter, we reported a ZnO nanowires metal-semiconductor-metal (MSM) Ultraviolet photodetector fabricated on glass substrate. High photocurrent to dark current ratio was achieved. Meanwhile, fast response and recovery time was observed for the device under 365 nm UV illumination. The mechanism of the photoresponse was discussed.

2. Experimental

ZnO nanowires used in this study were synthesized in horizontal tube furnace by chemical vapor deposition method. The source material, Zinc oxide powder(purity of 99.9%) and graphite powder with molar ratio of 1:1, was mixed and loaded onto a quartz boat placing at the center of the furnace. When the furnace was heated to $910\,^{\circ}\mathrm{C}$ an $\mathrm{Ar}:\mathrm{O_2}$ flow with a rate of 25/0.7 SCCM was introduced. The process was carried out for 30 min and white fluffy products were obtained on the substrate. The synthesized nanowires were characterized by scanning electron microscope (JSM-6490) and X-ray the diffraction (XRD, Rigaku DMAX-RB, Japan). The Photoluminescence (PL) measurements were carried out on a HITACHI 4500-type Vis-UV spectrophotometer with a Xe lamp as the excitation light source at room temperature.

The ultraviolet photodetector was processed to rectangle, and two parallel Pt electrodes were evaporated on the glass and ZnO nanowires were dispersed in ethonal and then applied to the gap between the electrodes. To characterize the photoresponse properties of the photodetector, a He-Cd (325 nm, 5 mW) laser was used as the excitation source and the response was measured by the Keithley 4200.

3. Results and Discussions

Morphology and crystal structure of the as-grown ZnO nanowires are shown in Fig. 1. A scanning electron microscopy (SEM) image of ZnO nanowires grown on the Si substrate is shown in Fig. 1(a). The ZnO nanowires are of uniform diameter, length, and density. The average length and diameter of the ZnO nanowires are about tens of micrometers and 100~200 nm, respectively. Fig. 1(b) shows the XRD pattern of ZnO nanowires grown on the Si substrate. The pattern clearly shows distinct peaks at 31.6, 34.2 and 36.1 corresponding to (100), (002) and (101) reflection. The high intensity peak at 36.1 corresponding to (002) plane indicates that the ZnO nanowires are highly oriented and crystalline in nature.

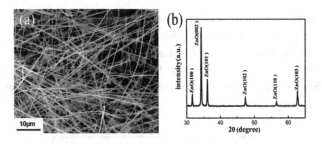

Figure 1. (a) SEM morphologies of ZnO nanowires, (b) XRD pattern of ZnO nanowires.

The PL spectrum of ZnO nanowires excited by a He-Cd (325 nm, 5 mW) laser is shown in Fig. 2. It can be seen that the ZnO nanowires have a sharp UV emission at 388 nm corresponding to the near band-edge emission is normally attributed to the exciton recombination [10]. And a broad weak green emission at 550~570 nm is related to the singly ionized oxygen vacancy because of recombination of a photogenerated hole with a single ionized electron in the valence band [11]. This characterization further confirms that the ZnO nanowires possesses high crystal quality.

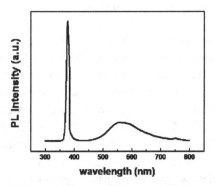

Figure 2. Room temperature PL spectrum of ZnO nanowires.

The current-voltage characteristics of the Pt-ZnO nanowires based metal-semiconductor-metal photodetector with and without UV illumination ($0.47mW/cm^2$, 365 nm) are shown in Fig. 3. The I-V curves were measured with bias from -50 V to 50 V at room temperature in ambient condition. It is worth noting that under the 5V bias, the photodetector showed a high photocurrent to dark current contrast ratio which was 9.44×10^3. The I-V curve of the Pt/ZnO nanowires PDs exhibited nonlinear in dark which is caused by the Schottky barriers (SB) formed between the semiconductor and the metal electrodes [12]. The I-V

126

curve of the PDs exhibited an asymmetrical shape in the positive and negative regions in dark because of the carriers trapping at the Pt/ZnO nanowires interface [13]. Meanwhile , the I-V curve tends to be linear and the slope increases dramatically under 365 nm UV illumination, which was reported elsewhere before. This is mainly because the decrease of the SB height between Pt/ZnO nanowires [14].

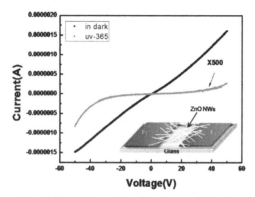

Figure 3. I-V characteristics of the ZnO nanowires photodetector in the dark and under illumination at UV 365 nm, inset is the schematic of the photodetector.

The photoresponse behavior of the device was characterized by measuring the current as a function of time when the 365 nm UV light was periodically turned on and off, as shown in Fig. 4. The measurement bias was fixed at 5V with UV light intensity at ~0.47 mW/cm^2. A fully reversible switching behavior was observed. The Pt/ZnO nanowires photodetector showed fast response and recovery time. Under the 5 V bias, the response (10%~90%) and recovery (90%~10%) time is 2 s and 1 s, which means the photodetector has a high sensitivity to 365 nm UV illumination even at a low voltage and indicats the Pt/ZnO nanowires photodetector could have a promising application in lower energy consumption detection.

When a ZnO nanowire is in ambient conditions, oxygen molecules are adsorbed onto its surface by capturing free electrons through $O_2(g)+e^- \rightarrow O_2^-(ad)$, creating a surface depletion layer thus reducing the conductivity. When illuminated with UV light with energy equal or higher than the bandgap, carrier density in the NW increases significantly. The positively charged hole neutralizes the adsorbed oxygen ions through $O_2^-(ad)+h^- \rightarrow O_2(g)$,thereby releasing the electron back to the conduction band and reducing in the depletion barrier thickness. Both the oxygen desorption and the increase in carrier density contributes to the conductance increase in the ZnO nanowire [15].

For Pt/ZnO nanowires photodetector, there is additional conducting mechanism that contributes to the UV response and recovery. The mechanism for the MSM structured Pt/ZnO nanowires photodetector can be attributed to the Schottky contact at the Pt/ZnO nanowires interfaces [16, 17]. Generally, the Pt (work function 5.65 eV) and n-type ZnO can form a Schottky contact. When the photodetector is exposed to 365 nm UV illumination, photogenerated electrons and hole in the SB interface region are separated by the strong electric field there. The electron-hole recombination rates decrease and the carrier lifetime increase and the SB height between semiconductor and the metal electrode decreases, resulting in an increase in free carrier density. However, the SB plays an important role in the high UV photo response and rapid response time.

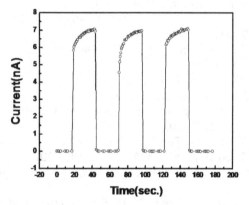

Figure 4. Spectral responsivity measurement of the photodetector upon UV (365 nm) illumination being turned on and off at voltage bias of 5V.

4. Conclusion

In conclusion, MSM UV photodetector based on Pt/ZnO nanowires provides a simple and low-cost way to fabricate high-performance UV photodetector. The photodetector showed high photocurrent to dark current ratio of 9.44×10^3 and fast response and recovery time under 365 nm UV illumination which can be attributed to the Schottky barrier between Pt/ZnO.

128

Acknowledgements

This work was supported by NSFC (51172022, 50972011), NSAF (10876001), the Research Fund of Co-construction Program from Beijing Municipal Commission of Education, the Fundamental Research Funds for the Central Universities. X. Q. Yan would like to thank the Beijing novel program (2008B19) and the Program for New Century Excellent Talents in University (NCET-09-0219).

References

1. L. Ying, F. Shiwei, Y. Ji, Z. Yuezong, X. Xuesong, *Optoelectr. China* **1**, 309(2008).
2. D. M. Bagnall, Y. F. Chen, Z. Zhu, T. Yao, S. Koyama, M. Y. Shen, T. Goto, *Appl. Phys. Lett.* **70**, 2230(1997).
3. D. C. Look, D. C. Reynolds, C. W. Litton, R. L. Jones, D. B. Eason, G. Cantwell, *Appl. Phys. Lett.* **81**, 1830(2002).
4. J. S. Young, W. L. Ji, H. T. Fang, J. S. Chang, K. Y. Su, L. X. Du, *Acta Mater.* **55**, 329 (2007).
5. J. S. Young, W. L. Ji, J. S. Chang, P. Y. Chen, M. S. Peng, *Semicond. Sci. Technol.* **23**, 085016(2008).
6. X. G. Zheng, Q. S. Li, J. P. Zhao, D. Chena, B. Zhao, Y. J. Yang, L. Ch. Zhang, *Appl. Surf. Sci.* **253**, 2264(2006).
7. N. N. Jandow, K. A. Ibrahim, H. A. Hassana, S. M. Thahab, O. S. S Hamad, *J. Electron. Dev.* **7**, 225(2010).
8. N. S. Liu, G. J. Fang, W. Zeng, H. Zhou, F. Cheng, Q. Zheng, L. Y. Yuan, X. Zou, X. Z. Zhao, *Appl. Mater. & Inter.* **2**, 1973(2010).
9. S. Hullavarad, N. Hullavarad, D. Look, B. Claflin, *Nanoscale Res. Lett.* **4**, 1421(2009).
10. K. Vanheusden, W. L. Warren, C. H. Seager, D. R. Tallant, J. A. Voigt, B. E. Gnade, *J. Appl. Phys.* **79**, 7983(1996).
11. B. J. Jin, S. H. Bae, Mater. *Sci. Eng. B.* **71**, 301 (2000).
12. Z. Y. Zhang, C. H. Jin, X. L. Liang, Q. Chen, L. M. Peng, *Appl. Phys. Lett.* **88**, 073102(2006).
13. J. S. Liu, C. X. Shan, B. H. Li, Z. Z. Zhang, C. L. Yang, D. Z. Shen, X. W. Fan, *Appl. Phys. Lett.* **97**, 251102(2010).
14. Y. Y. Li, X. Dong, C. W. Cheng, H. Q. Zhang, *Phys. B.* **404**, 4282(2009).
15. C. Soci, A. Zhang, B. Xiang, S. A. Dayeh, D. P. R. Aplin, J. Park, X. Y. Bao, Y. H. Lo, D. Wang, *Nano Lett.* **7**, 1003 (2007).
16. J. Zhou, Y. D. Gu, Y. F. Hu, W. J. Mai, P. H. Yeh, G. Bao, A. K. Sood, D. L. Polla, Z. L. Wang, *Appl. Phys. Lett.* **94**, 191103(2009).
17. Y. Jin, J. Wang, B. Sun, J. C. Blakesley, N. C. Greenham, *Nano Lett.* **8**, 1649(2008).

SOLUTION PROCESSED ZNO NANOROD ARRAYS/PFO HYBRID HETEROJUNCTION FOR LIGHT EMITTING

HANSHUO LIU, XIAOQIN YAN, XIANG CHEN, YOUSONG GU*

Department of Materials Physics and Chemistry, School of Materials Science and Engineering, University of Science and Technology Beijing, 30 Xueyuan Road, Beijing 100083, China

An organic-inorganic heterostructure light emitting diode consisting of the hole transporting layer poly (9,9-dioctylfluorenyl-2,7-diyl) (PFO) and n-type ZnO nanorod arrays is reported. ZnO nanorod arrays were synthesized by hydrothermal method, and characterized by scanning electron microscopy, X-ray diffraction, and photoluminescence spectrum. The current-voltage characteristics of the ZnO/PFO heterojunction showed diode behavior, and the turn-on voltage was 5.0 V. The light emitting phenomenon was observed in the heterostructure light emitting diode at the bias voltage of 7.0V.

1. Introduction

Considerable attention has been paid to the application of conjugated polymers to light emitting devices (LEDs) due to their potential use in flat panel [1-3] and solar cell. The main advantages of polymer-based devices are low power consumption, low cost, and simple process ability[4]. Zinc oxide (ZnO) is an II-VI semiconductor material with a wide band gap of about (3.37eV) together with a high exciton binding energy of (60 meV) at room temperature rendering ZnO to receive global attention especially in connection with the emerging nanotechnology paces toward functionality [5]. Moreover, ZnO possesses deep levels that emit light covering the whole visible spectrum [6]. ZnO nanostructures have gained substantial interest due to their simple fabrication routes, along with low cost and self-organization growth behavior enabling the growth of ZnO nanostructures on any substrate material regardless of lattice matching issues [7].Therefore, ZnO is a promising candidate for optoelectronic applications because of its suitable properties, such as high transparency, good electrical conductivity, tunable morphology, and large variety of possible nanostructures

---

*Corresponding author, E-mail:yousongu@mater.ustb.edu.cn

[8, 9].Based on the complementarities of charge transport properties, ZnO nanorod arrays(NRs)/polymer heterojunctions have been demonstrated for both solar cell devices and also in electroluminescence[10-12]. Likewise, using heterojunction strategy for photonic devices based on ZnO nanorod arrays has been a feasible way to obtain good performance LEDs [13-16].

In this letter, we reported that a heterostructure light emitting diode made of poly(9,9-dioctylfluorenyl-2,7-diyl) (PFO) layer and ZnO nanorod arrays fabricated by hydrothermal decomposition. The electroluminescence phenomenon was observed in the heterostructure LED at the bias voltage of 7V.

2. Experimental

ZnO nanorod arrays were synthesized on FTO glass, which was used as substrate and cleaned with acetone, ethanol and DI-water, respectively. The preparation of ZnO nanorod arrays was conducted through hydrothermal decomposition method at 95 ℃ . In brief, Zinc Nitrate Hexahydrate $(Zn(NO_3)_2 \cdot 6H_2O)$ (>99%) was dissolved in DI-water to form 0.05 M concentration, with 0.1 M of Hexamethylenetetramine(HMT, $C_6H_{12}N_4$) in 100 mL DI-water. The pre-coated substrates were transferred to the aqueous solution and kept for 12 hours in traditional laboratory oven. After that, a luminescent polymer (PFO) of 3mg/ml, was then spin coated on top of the ZnO nanorod arrays at spin speed of 3,500 rpm for 10 s. This coating process was applied three times. Subsequently, the 2.8 wt.% of PEDOT: PSS water emulsion used to increase hole injection was spin coated on the PFO layer and then annealed in vacuum furnace at 110℃. The last step in the fabrication of the hybrid LED was to utilize a metal contact to both parts of the hybrid junction. The Au film electrode of 50 nm thickness was achieved by thermal evaporation method on top of the fabricated device. Schematic illustrations of the hybrid LED is shown in Fig. 1.

The morphologies of the synthesized nanorods were characterized by scanning electron microscope (SEM, JSM-6490) and X-ray the diffraction (XRD, Rigaku DMAX-RB, Japan). The Photoluminescence (PL) measurements were carried out on a HITACHI 4500-type Vis–UV spectrophotometer with a Xe lamp as the excitation light source at room temperature. The current-voltage (I-V) characteristics were measured by Electrochemical Interfaces Workstation.

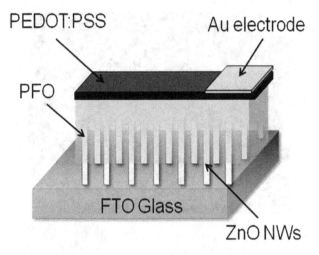

PEDOT:PSS   Au electrode

PFO

FTO Glass

ZnO NWs

Figure 1. Schematic diagram of the inorganic/organic heterostructure LED.

3. Results and Discussions

The morphological and structural characerics of ZnO nanorod arrays were obtained by SEM, and XRD, as shown in Fig. 2. The nanorods are straight and smooth in surface. The average diameter of the ZnO nanorods is ~100 nm and the length of the nanorods is ~3 μm. Fig. 3 shows the XRD pattern of ZnO nanorods. Three diffraction peaks can be observed, which indicates the ZnO wurtize strcture. The pattern clearly shows depicts distinct peaks at 31.6, 34.2 and 36.1 corresponding to (100), (002) and (101) reflections. The high intensity peak at 36.1 corresponding to the (002) reflection indicates that the ZnO nanorods are highly oriented and crystalline in nature.

The PL spectrum of ZnO nanorod arrays excited by a He-Cd (325 nm, 5 mW) laser is shown in Fig. 3. It can be seen that the ZnO nanorod arrays have a sharp UV emission at 388 nm corresponding to the near band-edge emission which is normally attributed to the exciton recombination [17]. And a broad much suppressed weak green emission at 550~570 nm is related to the singly ionized oxygen vacancy because of recombination of a photogenerated hole with a single ionized electron in the valence band [18]. It should be accepted that the ZnO nanorod arrays possesses high crystal quality, which caused the weak green emission is almost negligible.

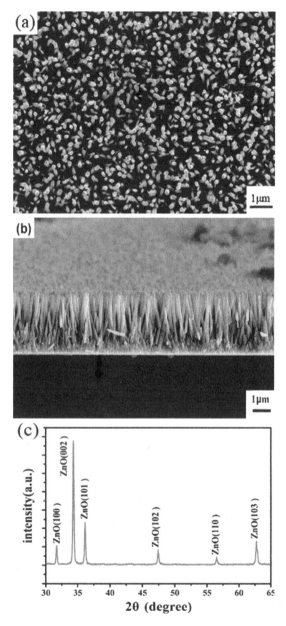

Figure 2. (a) and (b) SEM and FE-SEM morphologies of ZnO nanorod arrays. (c) XRD pattern of ZnO nanorods.

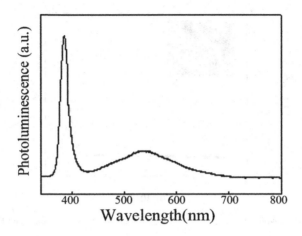

Figure 3. Room temperature PLspectrumof ZnO nanorod arrays.

Fig. 4 shows the current-voltage (I-V) characteristics of the fabricated hybrid light emitting diode consisting of FTO/ZnO/PFO/PEDOT:PSS/Au. The *I*-V characteristics show clear diode behavior, and the leakage current was found to be -7 μA at -5V. The general conclusion from the I-V curve is that this hybrid junction possesses the normal diode behavior. Due to energy band bending process the formation of sub bands take place and electrons and holes under forward bias voltages accumulate at the PFO/ZnO interface. The electrons existing at the interface between the LUMO level of the PFO and the conduction band edge of the ZnO nanorods continuously drop to the lower states and during the electron transitions continuous recombination between the electrons and holes happens leading to the emission of visible light. Conductive PEDOT:PSS was used as an hole injection layer. The injection of holes from PFO to ZnO takes place at 5 V, which defines the turn on voltage of the light emitting device. The device showed light emitting phenomenon when the voltage is up to7 V.

However, the phenomenon of luminescence disappeared with the voltage rise. This may be due to the insufficient contact of some area of ZnO/PFO interface.

134

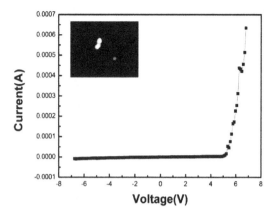

Figure 4. Current-voltage characteristics of the inorganic/organic heterostructure LED devices.

4. Conclusion

In summary, utilizing ZnO nanorod arrays synthesized by hydrothermal method and the PFO polymer, an inorganic-organic hybrid heterostructure LED with FTO/ZnO/PFO/PEDOT:PSS/Au structure were fabricated. The current-voltage characteristics reveal that the ZnO nanorod arrays/PFO hybrid heterojunction possesses ideal diode behavior, the turn-on voltage was 5 V. Electroluminescence was observed of the device at a voltage bias of 7 V. After this first demonstration, the next work is the device optimization that includes the choice of stable emitting polymers and the control of the PFO green emission to achieve white light emission.

Acknowledgements

This work was supported by NSFC (51172022， 50972011), NSAF (10876001), the Research Fund of Co-construction Program from Beijing Municipal Commission of Education, the Fundamental Research Funds for the Central Universities, the Beijing novel program (2008B19) and the Program for New Century Excellent Talents in University (NCET-09-0219).

Reference

1. J. H. Burroughes, D. D. C. Bradley, A. R. Brown, R. N. Marks, K. Mackay, R. H. Friend, O. L. Burns, A. B. Holmes, Nature. 347, 539 (1990).
2. D. Braun, A. J. Heeger, Appl. Phys. Lett. 58, 1982 (1991).
3. H. B. Wu, F. Huang, J. B. Peng, Y. Cao, Org. Electron. 6, 118 (2005).

4. A. Zainelabdin, S.Zaman, G. Amin, O. Nur, M.Willander, Nano. Res. Lett. 5, 1442 (2010).
5. Ü. Özgür, Y. I. Alivov, C. Liu, A. Teke, M. A. Reshchikov, S. Dogan, V. Avrutin, S. J. Cho, H. Morkoc, J. Appl. Phys. 98, 041301 (2005).
6. M. Willander, O. Nur, N. Bano, K. Sultana, New J. Phys. 11, 125020 (2009).
7. M. Willander, Q. X. Zhao, Q. H. Hu, P. Klason, V. Kuzmin, S. M. AlHilli, O. Nur, E. Lozovik, Superlattices Microstructures. 43, 352 (2008).
8. Z. L. Wang, X. Y. Kong, Y. Ding, P. Gao, L. Hughes, R. Yang, and Y. Zhang, Adv. Funct. Mater. 14, 943 (2004).
9. C. Ye, X. Fang, Y. Hao, X. Teng, L. Zhang, J. Phys. Chem. B 109, 19758 (2005).
10. C. D. Olson, J. Piris, T. R. Collins, E. S. Shaheen, S. D. Ginley, Thin Solid Films. 496, 26 (2006).
11. J. Sun, N. B. Pal, J. B. Jung, E. H. Katz, Org. Electron. 10, 1 (2009).
12. A. Wadeasa, O. Nur, M. Willander, Nanotechnology 20, 065710 (2008).
13. A. Wadeasa, O. Nur, M. Willander, Nanotechnology 20, 065710 (2009).
14. M. Willander, L. L. Yang, A. Wadeasa, S. U. Ali, M. H. Asif, Q. X. Zhao, O. Nur, J. Mater. Chem. 19, 1006 (2009).
15. A. Wadeasa, S. L. Beegum, S. Raja, O. Nur, M. Willander, Appl. Phys. A 95, 807 (2009).
16. A. Wadeasa, G. Tzamalis, P. Sehati, O. Nur, M. Fahlman, M. Willander, M. Berggren, X. Crispin, Chem. Phys. Lett. 490, 200 (2010).
17. K. Vanheusden, W. L. Warren, C. H. Seager, D. R. Tallant, J. A. Voigt, B. E. Gnade, J. Appl. Phys. 79, 7983 (1996).
18. B. J. Jin, S. H. Bae, Mater. Sci. Eng. B. 71, 301 (2000).

FABRICATION AND PERFORMANCE STUDY ON INDIVIDUAL ZNO NANOWIRES BASED BIOELECTRODE

YANGUANG ZHAO, XIAOQIN YAN[†]

State Key Laboratory for Advanced Metals and Materials, School of Materials Science and Engineering, University of Science and Technology Beijing, 30 Xueyuan Road, Beijing 100083, People's Republic of China

ZHUO KANG, PEI LIN

Key Laboratory of New Energy Materials and Technologies, University of Science and Technology Beijing, Beijing 100083, People's Republic of China

One-dimensional zinc oxide nanowires (ZnO NWs) have unique advantages for use in biosensors as follows: oxide stable surface, excellent biosafety, high specific surface area, high isoelectric point (IEP = 9.5). In this work, we have prepared a kind of electrochemical bioelectrode based on individual ZnO NWs. Here, ZnO NWs with high quality were successfully synthesized by CVD method, which were characterized by scanning electron microscopy, X-ray diffraction and photoluminescence. Then the Raman spectra and electrical characterization demonstrated the adsorption of uricase on ZnO wires. At last, a series of electrochemical measurements were carried out by using an electrochemical workstation with a conventional three-electrode system to obtain the cyclic voltammetry characteristics of the bioelectrodes. The excellent performance of the fabricated bioelectrode implies the potential application for single ZnO nanowire to construct electrochemical biosensor for the detection of uric acid.

1. Introduction

Nowadays, there is an increasing demand of highly-sensitive sensing devices having potential applications in biomedical, biotechnological, industrial and environmental fields, especially disease diagnosis [1]. Consequently, over the past few decades, various biosensors, especially electrochemical devices have traditionally attracted great attention in biosensor development [2,3]. Electrochemical biosensor is a kind of molecular sensing device which intimately couples a biological recognition element to an electrode transducer.

---

† Corresponding author, e-mail: xqyan@mater.ustb.edu.cn

One-dimensional (1-D) nanomaterials, such as silicon nanowires [4,5], In_2O_3 nanowires [6], ZnO nanowires [7], ZnO nanorod [8], ZnO nanorod arrays [9,10], ZnO nanocomb [11], carbon nanotube [12] are particularly attractive for bioelectronic detection. Among those 1-D nanomaterials, ZnO nanowires [13] have attracted more attention because of their nontoxicity, excellent biological safety, biocompatibility, high isoelectric point (IEP = 9.5) and fine electron transport property.

In our work, ZnO NWs with high quality were successfully synthesized using a chemical vapor phase transport method and characterized by SEM, XRD and PL. Then individual ZnO NWs were modified by uricase to fabricate bioelectrode and the adsorption of uricase on ZnO wires was demonstrated by Raman spectra. Furthermore, the electrical properties and cyclic voltammetry curves of different electrodes were systematically investigated and compared. The Raman spectra demonstrate that the uricase was successfully immobilized on the individual ZnO NWs, and the performance test of different electrodes shows that the individual ZnO NWs has the potential application to construct biosensor for the detection of uric acid.

2. Experimental details

ZnO nanowires were prepared in tubular furnace by using a vapor phase transport method. The mixture of ZnO powder of high purity (99.99%) and carbon powder with the molar ratio of 1:1.2 was placed into a small Al_2O_3 boat as the evaporation source. A slice of silicon wafer was then fixed on the top of the quartz boat facing the mixed powder. The furnace was heated to 980 $^\circ C$ and then the quartz boat was placed in the middle of the horizontal quartz tube. The Ar/O_2 with a flow rate of 297/3 sccm was introduced for 20 minutes and a white flocky product was found on the Si substrate. The prepared products were characterized by scanning electron microscopy (SEM JEOL-6490), X-ray diffraction (XRD, Rigaku DMAX-RB, Japan), respectively. The PL spectrum was obtained by utilizing the confocal Raman spectroscopy (JY-HR800) at room temperature, with a He-Cd laser as the excitation source.

The Raman spectrum and electrical property of the single ZnO nanowire before and after the immobilization of uricase were obtained by utilizing the confocal Raman spectroscopy (JY-HR800) with an Ar^+ laser source at room temperature and Keithley 4200 SCS (Agilent, USA) combining with probe station (PW-600), respectively.

Furthermore, various columnar electrodes were fabricated and the cyclic voltammetry characteristics of them under different conditions were performed

138

by utilizing an electrochemical workstation (Solartron SI 1287) with a conventional three-electrode system.

3. Results and discussion

Fig. 1(a) shows a typical scanning electron microscope (SEM) image of the prepared ZnO nanowires. It's seen that the ZnO NWs were with an average diameter of 800 nm and a length of 25 μm or so. To characterize the overall structure of the ZnO NWs, the XRD pattern of the samples was obtained. As is shown in Fig. 1(b), all the peaks match the standard wurtzite structure of ZnO with the lattice constants of a = 0.3250 nm and c = 0.5207 nm, and no diffraction peak of impurity phase is found. Fig. 1(c) illustrates the PL spectrum of the ZnO NWs. Only a strong peak at 381 nm corresponding to the near-band-edge emission (NBE) from free exciton is observed, implying the high crystal quality of the products [14]. Fig. 1(d) depicts the SEM image of the individual ZnO NW fixed by the deposition of platinum (Pt) source utilizing FIB system.

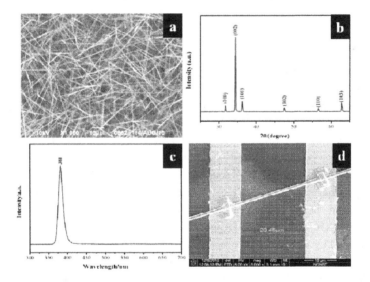

Fig. 1. (a) Top-view SEM image of ZnO nanowires grown on Si substrate. (b) XRD pattern of ZnO nanowires. (c) PL spectrum of ZnO nanowires. (d) Individual ZnO nanowire fixed to Au contact leads on the SiO_2/Si wafer.

To investigate the structure and crystallization of ZnO nanowires further, the Raman measurements were carried out. The typical Raman spectrum of ZnO is shown in Fig. 2 by red curve. The peaks at 98, 330, 378, 437, 578 and

1150 cm^{-1} are assigned to E$_2^{low}$, E$_2^{high}$-E$_2^{low}$, A1(TO), E$_2^{high}$, E1(LO) and 2E1(LO) modes of ZnO [15], respectively. The Raman peak of uricase on the blue curve is observed at 980 cm^{-1} and the spectrum of ZnO nanowire modified by uricase is illustrated by black curve. As is seen, the peaks of the Raman spectrum well coincide with those of pure ZnO nanowire and uricase, which demonstrates that the immobilization of uricase on ZnO NWs didn't change the structure of ZnO at all and well maintained the structure and activity of uricase as well.

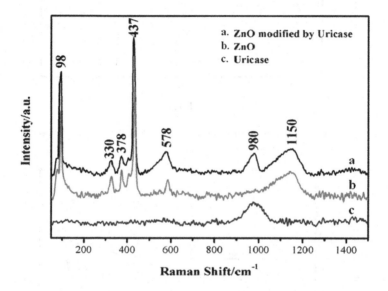

Fig. 2. Raman Spectra of ZnO, uricase and ZnO modified by uricase

The electrical characteristics of the individual ZnO nanowire before and after the immobilization of uricase were also compared and investigated, as shown in Fig. 3. It's found that the current of the ZnO nanowire decreases after the immobilization of uricase. The surface of ZnO with high isoelectric point (IEP, 9.5) and uricase (IEP = 5.4) are positively and negatively charged under our working condition (PBS with a pH of 7.4), respectively. Hence, the immobilization of uricase on the ZnO nanowire is highly facilitated through the electrostatic interaction between them. The macromolecular coating and immobilization of uricase lead to the carrier scattering and cause the decrease of the conductivity of ZnO nanowire [16].

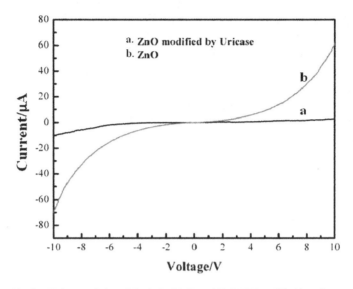

Fig. 3. I-V characteristics of single ZnO NW and ZnO NW modified by uricase

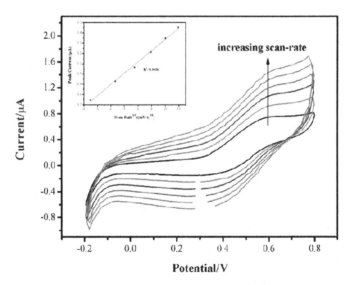

Fig. 4. CV curves of the Nafion/uricase/ZnO/Au electrode at different scan rates in PBS with 0.5 mM uric acid and the schematic diagram (inset) of peak current at various scan rates versus the square root of scan rates

Furthermore, a series of electrochemical measurements were carried out to investigate the properties of the conductive bioelectrodes. The effect of scan rate on the electrochemical response of the fabricated electrode is evaluated firstly. The CV sweep curves of the Nafion/uricase/ZnO/Au electrode in the PBS (0.01M, PH 7.4) with 0.5 mM uric acid at various scan rates from 20 to 120 mV s^{-1} are presented in Fig. 4 and the inset demonstrates the fitted curve of peak currents versus scan rates. It's shown that the anodic peak current is linearly proportional to the square root of scan rate, indicating a typical diffusion-controlled electrochemical behavior.

To evaluate the effect of the individual ZnO nanowire on the performance of the fabricated bioelectrode, the electrocatalytic activities of three different electrodes in 0.01M PBS with a PH of 7.4 at the scan rate of 50mV s^{-1} were investigated, as shown in Fig. 5. It is demonstrated that the current response of the Nafion/Au electrode and the Nafion/uricase/Au electrode in the PBS are similar to each other and no redox peak is observed in both of the CV curves. However, the current value of the Nafion/uricase/ZnO/Au electrode increases significantly in contrast to the above two and there are weak redox peaks at approximately 0.55V and 0.31V, respectively, suggesting that ZnO nanowire has a favorable influence on the performance of the bioelectrode.

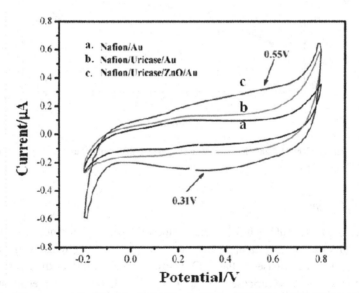

Fig. 5. CV curves of different electrodes in PBS: Nafion/Au, Nafion/uricase/Au and Nafion/uricase/ZnO/Au

Fig. 6 illustrates the CV characteristics of the Nafion/uricase/ZnO/Au electrode in the PBS (0.01M, pH = 7.4) with uric acid solutions of different concentrations (0, 0.05, 0.1, 0.2, 0.4 and 1mM). The oxidation current enhances significantly with the increment in the concentration of the uric acid solutions, suggesting the relationship existing between the current value and the uric acid concentration. A strong oxidation peak appears at about 0.59 V on the CV curve in the PBS with the addition of 1mM uric acid, which can be ascribed to the H_2O_2 produced during the uric acid oxidation catalyzed by uricase and the electrons generated from the redox reaction of H_2O_2 through the electrode have enhanced the current response.

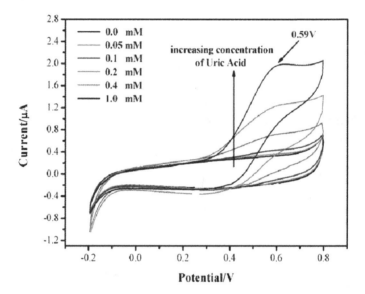

Fig. 6. CV curves of the Nafion/uricase/ZnO/Au electrode in PBS without uric acid and in PBS with uric acid of various concentrations

4. Conclusions

In conclusion, the ZnO nanowires fabricated by the CVD method were of high quality. The Raman spectra and electrical characterization demonstrated that the immobilization of uricase on the ZnO nanowire can well retain the structure of ZnO and the activity of uricase. A series of cyclic voltammetry curves illustrated that the individual ZnO NW played an important role of electronic transmission channel for the bioelectrode and the current response enhanced with the increment of the uric acid concentration. All the results have implied the potential application for the fabricated bioelectrode to construct biosensor based on individual ZnO nanowire for the detection of uric acid.

Acknowledgments

This work was supported by NSFC (51172022, 50972011), NSAF (10876001), the Research Fund of Co-construction Program from Beijing Municipal Commission of Education, the Fundamental Research Funds for the Central Universities and the Beijing novel program (2008B19).

References

1. S.L. Brooks, R.E. Ashy, A.P.F. Turner, M.R. Calder and D.J. Clarke, *Biosensor* **3**, 45 (1987).
2. Turner, A.P., Karube, I., Wilson, G., *Biosensors: fundamentals and applications*. Oxford Science Publications, Oxford (1986).
3. Wang, J., *Analytical Electrochemistry*. Wiley, New York, 2000a.
4. Xihua Wang, Yu Chen, Katherine A. Gibney, etc., *Appl. Phys. Lett.* **92**, 013903 (2008).
5. Xuan P. A. Gao, Gengfeng Zheng and Charles M. Lieber, *Nano Lett.* **10**, 547 (2010).
6. C. Li, M. Curreli, H. Lin, B. Lei, F. N. Ishikawa, R. Datar, R. J. Cote, M. E. Thompson, C. Zhou, *J. Am. Chem. Soc.* **127**, 12484 (2005).
7. Ahmi Choi, Kyoungwon Kim, Hyo-Il Jung, Sang Yeol Lee, *Sens. Actuators.* **B148**, 57 (2010).
8. Jin Suk Kim, Won Il Park, Chul-Ho Lee and Gyu-Chul Yi, *J. Korean Phys. Soc.* **49**, 1 (2006).
9. Wei A, Sun X W, Wang J X, et al., *Appl. Phys. Lett.* **89**, 123902-1 (2006).
10. Kang B S, Wang H T, Ren F, et al., *Appl. Phys. Lett.* **91**, 252103-1 (2007).
11. Wang J X, Sun X W, Wei A, et al., *Appl. Phys. Lett.* **88**, 233106-1 (2006).
12. Masuhiro Abe, Katsuyuki Murata, Atsuhiko Kojima, Yasuo Ifuku, Mitsuaki Shimizu, Tatsuaki Ataka and Kazuhiko Matsumoto, *J. Phys. Chem.* **C111**, 8667 (2007).
13. Jun Zhou, Ningsheng Xu and Zhong L. Wang, *Adv. Mater.* **18**, 2432 (2006).
14. Y. C. Kong, D. P. Yu, B. Zhang, W. Fang and S. Q. Feng, *Appl. Phys. Lett.* **78**, 407 (2001).
15. T.C. Damen, S.P.S. Porto, B. Tell, *Phys. Rev.* **142**, 570 (1966).
16. Cristian Staii and Alan T. Johnson, Jr., *Nano. Lett.* **5**, 1774 (2005).

ENZYME-BASED LACTIC ACID DETECTION USING ALGAAS/GAAS HIGH ELECTRON MOBILITY TRANSISTOR WITH SB-DOPED ZNO NANOWIRES GROWN ON THE GATE REGION

SIWEI MA, YUNHUA HUANG*, HANSHUO LIU, XIAOHUI ZHANG, QINGLIANG LIAO

Department of Materials Physics and Chemistry and State Key Laboratory for Advanced Metals and Materials, University of Science and Technology Beijing, 30 Xueyuan Road, Beijing 100083, China

Sb-doped ZnO nanowires were synthesized via chemical vapor deposition method. Scanning electron microscopic, transmission electron microscopic, X-ray diffraction and energy dispersive spectrometer have been used to characterize the morphology and structure of the nanowires. The AlGaAs/GaAs HEMT drain-source current exhibited a fast response of about 1s when different concentrations of lactic acid solutions were added to the surface of lactate oxidase immobilized on the ZnO nanowires. The HEMT could detect a range of lactic acid concentrations from 3 pM to 30 μM. The biosensor exhibited good performance along with fast response, high sensitivity, and long-term stability. Our results demonstrate the possibility of using AlGaAs/GaAs HEMTs for lactic acid measurements and provide new further fundamental insights into the study of nanoscience and nanodevices.

1. Introduction

Considerable attention has been paid to developing improved methods for detecting lactic acid due to its importance in medicine and food chemistry. Since elevated concentrations of lactate may not only indicate the presence but also the severity of these clinically important disorders, accurate measurements of blood lactate concentration is critical to their proper management. For example, lactic acid concentrations can be used to monitor the physical condition of athletes or of patients with chronic diseases such as heart failure, diabetes, and chronic renal failure [1]. Moreover, in the food industry, lactic acid levels serve as an indicator of freshness, stability, and storage quality [2]. For the reasons stated above, it is desirable to develop an excellent lactic acid biosensor capable of

---

* Corresponding author's E-mail: huangyh@mater.ustb.edu.cn

simple and real-time measurements, rapid response, wide range of detection concentration, high specificity and sensitivity, and low price.

Recently, the high electron mobility transistors (HEMTs) have shown promise for a variety of chemical and biological sensing applications [3~8]. Nowadays, ZnO nanorod-gated AlGaN/GaN HEMTs have been used to detect glucose with a detection limit of 0.5 nM [9] and detect lactic acid with a wide range of concentrations, ranging from 167 nM to 139 μM [2]. The favorable electrostatic interaction between ZnO and lactate oxidase (LOX) was achieved due to their huge difference in isoelectric points (IEPs) [10, 11]. In addition, Nanowires with the unique catalytic activities and biocompatibility are widely used to improve the performance of electrochemical biosensors, such as decrease in overpotential, direct electron transfer capabilities, and high sensitivity and selectivity [12~16].

In this paper, we firstly realized the application of Sb-doped ZnO nanowire in real-time detection of lactic acid. In comparison with pure ZnO nanowire, the Sb-doped ZnO nanowire displayed excellent electron transfer properties for the enzymes and better ability to retain the electroactivity of enzymes [17]. The Sb-doped ZnO nanowires-gated HEMT biosensor was constructed and exhibited good performances along with rapid response, low detection limit, high sensitivity, and long-term stability. Our study reveals that Sb-doped ZnO nanowire is promising platform for the construction of mediator-free biosensors and portable, fast response, and wireless-based lactic acid detectors can be realized with AlGaAs/GaAs HEMT based biosensors.

2. Experimental

The Sb-doped ZnO nanowires were synthesis through the mixture of zinc, Sb_2O_3 and graphite powders with the mole ratio of 30:1:2 in an Al_2O_3 boat inside a quartz tube at 850 °C, and the pre-cleaned silicon substrate coated with 30 nm Au nanopaticles was fixed on the boat. Then the boat was placed at the center of the horizontal tube furnace. Ar was used as the carrier gas and O_2 as the reaction gas. Ar and O_2 were flowed constantly at a rate of 300 standard cm^3min^{-1} (SCCM) and 3 standard cm^3min^{-1} (SCCM), respectively. The growth time is about 30 minutes. After the reaction, white color product was obtained on the Au coated area of the substrates.

The morphologies of the products were characterized by field emission scanning electric microscopy (FE-SEM) (Zeiss, SUPRA-55). Energy dispersive X-ray spectra (EDS) (SUPRA-55), X-ray diffraction (XRD) (Rigaku, DMAX-RB), and transmission electron microscope (TEM) (JEOL-2010), were designed to examine the composition, structures and magnetic properties of the Sb-doped ZnO nanowires.

The lactate oxidase solution was prepared with concentration of 1.3 mg/ml in 250 μM phosphate buffer saline (pH value of 7.4). The gate area of the HEMTs was modified by the Sb-doped ZnO nanowires prepared above. Before fabrication the device, the nanowires were dissolved sufficiently in the ethanol solution by ultrasonic device. Using a series of pipettes, 15 μL the nanowires suspension, was exactly transported to the gate area. Subsequently, 3 μL lactate oxidase (~100U/mg, Sigma Aldrich) solution, at a concentration of 5 units/ml, was introduced to the former nanowires dried at room temperature. It is the traditional physical adsorption method that can avoid reducing the enzyme activity and the procedure is more convenient than other enzyme immobilization methods. For protecting the nanowires and lactate oxidase, 3 μL Nafion (Sigma Aldrich) solutions were also added to the gate area, covering the nanowires. In order to immobilize the nanowires on the gate area, the HEMTs was then put in an opaque and sealed box to form a film of Nafion under a condition without light for 15 minutes. After that, the device was put in a refrigerator at 4 ℃ for 24 hours for lactate oxidase (LOX) immobilization and the lactate biosensor had been fabricated well for detecting the concentration of lactic acid.

3. Results and Discussions

The composition, morphological and structural characters of Sb-doped nanowires were obtained by SEM, TEM, XRD and EDS analysis, as shown in Figure 1. The SEM image is given in Figure 1 (a) and the inset is the FE-SEM image. The nanowires have the diameters in the range of 200-300 nm, and the average length of about 30 μm. Figure 1 (b) and the inset shows the TEM images and HRTEM images of the Sb-doped ZnO nanowires, which confirm that the Sb-doped ZnO nanowires are single-crystalline phases. The XRD spectrum of Sb-doped ZnO nanowires is shown in Figure 1 (c). All reflections of the nanowires are in excellent accordance with a hexagonal wurtzite structure. The diffraction peaks are similar to those of pure ZnO nanowires. No peaks of other impurities, such as Sb or Sb_2O_3, are detected. The diffraction peaks also indicated that the nanowires were Sb-doped ZnO with single phase. The EDS spectrum of Sb-doped ZnO nanowires in Figure 1 (d) confirms that the composition of the nanowires is Zn, O and Sb, and the Sb content is about 6.0 atom %.

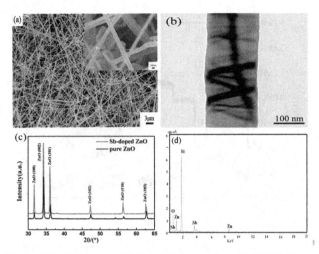

Figure 1. (a) SEM and FE-SEM image of Sb-doped ZnO nanowires; (b) HRTEM image of Sb-doped ZnO nanowires; (c) XRD pattern of Sb-doped and pure ZnO nanowires; (d) Energy dispersive spectrum of Sb-doped ZnO nanowires.

The real time lactic acid detection in PBS buffer solution using the drain current change with a constant bias of 500 mV was showed in Figure 2. The HEMT biosensor was first exposed to PBS and no change of the drain current was detected with the addition of buffer solution at around 400 s, showing the stability of the device. This stability was important to exclude possible noise from the mechanical change of the lactic acid solution. In clear contrast, the current change showed a rapid response of less than 1 s when the target lactic solution was switched to the surface of the device. The currents change due to the exposure of lactic acid molecules in a buffer solution was stabilized after the lactic acid molecules thoroughly absorbed onto the surface of nanowires. Different concentrations (from 3 pM to 300 mM) of the exposed target lactic acid in a buffer solution were detected and the limit of detection of this sensor was 3 pM.

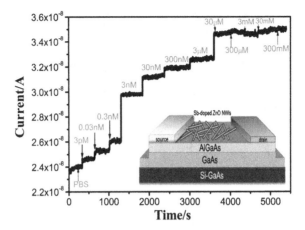

Figure 2. Plot of drain current vs time with successive exposure to higher lactic acid concentrations, ranging from 3 pM to 300 mM in PBS with a pH value of 7.4.

The lactic acid sensing with the HEMT sensor was measured through the drain current of HEMT with a change in the charges on the ZnO nanowires and the detection signal was amplified through the HEMT device. A significant change in drain current was exihibited for exposing the HEMT to the lactic acid at a low concentration of 3 pM due to this amplification effect. In addition, the amount of sample only depended on the area of gate dimension, can be minimized for the HEMT sensor due to the fact that no reference electrode is required. The excellent sensitivity of the HEMT sensor reported herein coupled to its low sample volume requirement makes this approach ideal for various medical applications, such as measuring lactic acid levels in exhaled breath condensate as a surrogate for blood-based lactate measurements [18]. Figure 3 shows the changes of drain current as a function of lactic acid concentration and each data point is based the average of five different measurements. The linear relationship of different lactic acid concentration versus drain current change was examined.

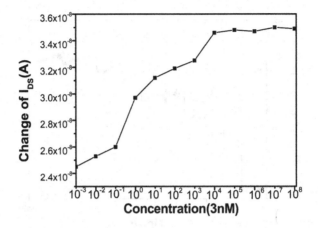

Figure 3. Change of drain current in HEMT sensor vs different concentrations from 3 pM to 300 mM with a *p*H value of 7.4.

For investigating the importance of the Sb-doped ZnO nanowires in HEMT biosensor, the performance of no ZnO nanowires-gated HEMT was investigated and the result of detection is shown in Figure 4. The signal of the response current was irregular and the response time of every lactic acid concentration was postponed to 10 s. Meanwhile, some of the concentrations could not be detected accurately when added the lactic acid solution of different concentration (from 3 pM to 300 mM) onto the gate region of HEMT sensor. The test result demonstrates that the ZnO nanowires could absorb and immobilize LOX efficiently and afford a compatible micro-environment for retaining the activity of LOX. Therefore, the Sb-doped ZnO nanowires play a critical role in improving the property of HEMT sensor.

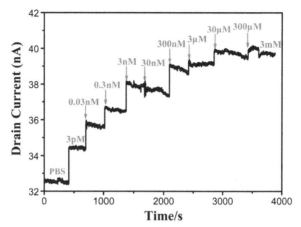

Figure 4. Drain current of no ZnO naonowires-gated AlGaAs/GaAs HEMT vs time for lactic acid from 3 pM to 3 mM with a pH value of 7.4.

4. Conclusion

In summary, this work has a new insight into the effect of doping of semiconductor nanowires on the electrical properties of immobilized LOX, and has shown that the Sb-doped ZnO nanowire-gated region of an AlGaAs/GaAs HEMT structure can be functionalized with LOX for the detection of lactic acid with a limit of detection of 3 pM. Moreover, the Sb-doped ZnO nanowire modified HEMT biosensor exhibited fast response, low detection limit, high sensitivity and excellent stability for the detection of lactic acid. This electronic detection of biomolecules is a significant step toward a compact sensor integrated with commercial available wireless transmitter to realize a real-time, fast response, and high sensitivity lactic acid detection.

Acknowledgements

This work was supported by the Major Project of International Cooperation and Exchanges (2006DFB51000), NSFC (51172022), NSAF (10876001), the Research Fund of Co-construction Program from Beijing Municipal Commission of Education, the Fundamental Research Funds for the Central Universities.

References

1. B. Phypers, T. Pierce, *Critical Care & Pain* 6. **128,** (2006).
2. B. H. Chu, B. S. Kang, F. Ren, C. Y. Chang, Y. L. Wang, S. J. Pearton, A. V. Glushakov, D. M. Dennis, J. W. Johnson, P. Rajagopal, J. C. Roberts, E. L. Piner, K. J. Linthicum, *Appl. Phys. Lett.* **93,** 042114 (2008).
3. A. P. Zhang, L. B. Rowland, E. B. Kaminsky, V. Tilak, J. C. Grande, J. Teetsov, A. Vertiatchikh, L. F. Eastman, *J. Electron, Mater.* **32,** 388 (2003).
4. B. S. Kang, F. Ren, L. Wang, C. Lofton, W. Tan, S. J. Pearton, A. Dabiran, A. Osinsky, P. P. Chow, *Appl. Phys. Lett.* **87,** 023508 (2005).
5. R. Neuberger, G. Muller, O. Ambacher, M. Stutzmann, *Phys. Status Solidi A.* **185,** 85 (2001).
6. J. Schalwig, G. Muller, O. Ambacher, M. Stutzmann, Phys. *Status Solidi A.* **185,** 39 (2001).
7. G. Steinhoff, M. Hermann, W. J. Schaff, L. F. Eastman, M. Stutzmann, M. Eickhoff, *Appl. Phys. Lett.* **83,** 177 (2003).
8. M. Eickhoff, R. Neuberger, G. Steinhoff, O. Ambacher, G. Muller, M. Stutzmann, *Phys. Status Solidi B* **228,** 519 (2001).
9. B. S. Kang, H. T. Wang, F. Ren, S. J. Pearton, L. T. E. Morey, D. M. Dennis, J. W. Johnson, P. Rajagopal, J. C. Roberts, E. L. Piner, K. J. Linthicum, *Appl. Phys. Lett.* **91,** 252103 (2007).
10. J. Di, J. Cheng, Q. Xu, H. Zheng, J. Zhuang, Y. Sun, K. Wang, X. Mo, S. Bi, *Biosens. Bioelectron.* **23,** 682 (2007).
11. A. Lupu, A. Valsesia, F. Bretagnol, P. Colpo, F. Rossi, Sens. *Actuators* B.**127,** 606 (2007).
12. A. Liu, M. D. Wei, I. Honma, H. Zhou, Adv. *Funct. Mater.* **16,** 371 (2006).
13. J. P. Liu, C. X. Guo, C. M. Li, Y. Y. Li, Q. B. Chi, X. T. Huang, L. Liao, T. Yu, *Electrochem. Commun.* **11,** 202 (2009).
14. Y. S. Lu, M. H. Yang, F. L. Qu, G. L. Shen, R. Q. Yu, *Bioelectrochemistry,* **71,** 211 (2007).
15. J. Wang, M. Musameh, *Anal. Clim. Acta.* **539,** 209 (2005).
16. X. Y. Zhang, D. Li, L. Bourgeois, H. T. Wang, P. A. Webley, *Chem. Phys. Chem.* **10,** 436 (2008).
17. L. Li, J. Huang, T. H. Wang, H. Zhang, Y. Liu, J. H. Li, *Biosen. Bioelectron.* **25,** 2436 (2010).
18. B. S. Kang, H. T. Wang, F. Ren, S. J. Pearton, L. T. E. Morey, D. M. Dennis, J. W. Johnson, P. Rajagopal, J. C. Roberts, E. L. Piner, K. J. Linthicum, *Appl. Phys. Lett.* **91,** 252103 (2007).

SINGLE ZNO NANOWIRE-BASED BIOFET SENSORS FOR ULTRASENSITIVE, LABEL-FREE AND REAL-TIME DETECTION OF URIC ACID

PEI LIN, XI LIU, XIAOQIN YAN*, ZHUO KANG, YANG LEI, YANGUANG ZHAO

*State Key Laboratory for Advanced Metals and Materials, School of Materials Science and Engineering, University of Science and Technology Beijing, Beijing 100083, People's Republic of China
and Key Laboratory of New Energy Materials and Technologies, University of Science and Technology Beijing, Beijing 100083, People's Republic of China*

Qualitative and quantitative detection of biological and chemical species is crucial in many areas, ranging from clinical diagnosis to homeland security. Due to the advantages of ultrahigh sensitivity, label-free, fast readout and easy fabrication over the traditional detection systems, semiconductor nanowire based electronic devices have emerged as a potential platform. In this paper, we fabricated a single ZnO nanowire-based bioFET sensor for the detection of low and high concentration uric acid solution at the same time. The addition of uric acid with the concentrations from 1 pM to 0.5 mM resulted in the electrical conductance changes of up to 227 nS, and the response time turns out to be in the order of millisecond. The ZnO NW biosensor could easily detect as low as 1 pM of the uric acid with 14.7 nS of conductance increase, which implied that the sensitivity of the biosensor can be below the 1pM concentration.

1. Introduction

Currently, fabrication of nanoscale glucose sensors based on one-dimensional nanomaterials such as Si [1], In_2O_3 [2], ZnO nanowires [3], Carbon nanotubes [4], graphene [5] have attracted much attention. In comparison to planar and bulk materials, the one-dimensional morphology and nanometer-scale cross-section of nanowires lead to depletion or accumulation of carriers in the 'bulk' of the device when a charged species binds to the surface [6~10]. Based on the specific feature of nanowire, the detection level of biosensor may go down to the ultimate level of a single molecule. What's more, the detection is monitored in terms of significant change in electrical properties which eliminates the need of labor-intensive labelling and complex measurement equipment. The fundamental principle of the FET biosensor relies on its sensitive response to the

---

* E-mail: xqyan@mater.ustb.edu.cn

variation of electric field or potential at the surface resulting from the binding of charged molecules [8, 11, 12]. Most of the current works are focused on silicon nanowire and carbon nanotube, and only a limited number of studies describe the detection of bio species using oxide semiconductor. However, the unstable silicon nanowire surfaces can be easily oxidized and form an insulating layer which may degrade the device reliability and sensitivity, while the chirality of carbon nanotube remains an unsolved problem.

Among the oxide semiconductor, ZnO nanowires [13] are believed to be the best candidate as the future integrated biosensors because of its excellent electrical properties and biocompatibility. Through proper thermal annealing and surface passivation, the ZnO nanowire based FET could exhibit high electron mobility above 1000 cm^2/Vs. Meanwhile, the ZnO nanowires have active surfaces that can be easily modified for the immobilization of numerous biomolecules [14~15]. Most of all, the ZnO nanowire based FET biosensor can be integrated with Si-based signal processing and communication circuits. These notable advantages over other non-oxide semiconductors make ZnO a promising material for nano-scaled biosensors.

Uric acid is created when the body breaks down purine nucleotides. High blood concentration of uric acid can lead to a type of arthritis known as gout. Also, the monitoring of urea concentration in blood is a way to evaluate kidney disease. When uric acid reaches the enzyme-functionalized surface, the uricase catalyzes the following reaction [16]

$$Uric\ Acid + O_2 \xrightarrow{Uricase} Allantoin + H_2O_2 + CO_2$$

$$H_2O_2 \longrightarrow O_2 + 2H^+ + 2e^-$$

Here, we report the design and construction of single ZnO nanowire based FET biosensor to detect uric acid concentration. Firstly, uniform and single crystal nanowires were obtained through CVD method. Secondly, using the covalent modification method, uricase was linked to the surface closely and retained its full pristine activity. Thirdly, current-voltage (I_{ds}-V_{ds}) was used to monitor the change in the conductance during the successive addition of different concentrations of uric acid. The results demonstrate the applications of ZnO nanowire devices for label-free, ultrasensitive and real-time detection of a wide range of biological and chemical species.

2. Materials and methods

2.1. *ZnO nanowires synthesis*

Ultralong ZnO nanowires were synthesized in CVD furnace without any catalysts. The mixture of ZnO and carbon powder with the molar ratio of 1:1 was initially uniformly grinded, and a suitable amount of powder was placed in an Al_2O_3 boat inside a quartz tube as the evaporation source. A silicon substrate without any catalysts was then fixed on the top of the source boat to collect the as-synthesized nanowires. Ar was used as the carrier gas, and O_2 was the reaction gas. Then the whole system was maintained at the temperature of 980 ℃ with the flow rate of Ar/O_2 at 297/3 sccm (standard-state cubic centimeter per minute) for 20 minutes. After the reaction, the furnace was cooled in Ar for about 1h and then the substrate with a white flocky product was taken out.

2.2. *Fabrication of Single ZnO NW-based bioFET*

ZnO NW-based FET biosensors were fabricated for the real time detection of biomolecular interactions. We adopted the widely used fabrication methods reported by other groups for NW FET fabrication [4, 17]. The high quality ZnO NW synthesized on silicon wafer was aligned by probe station. A highly doped n-type Si wafer was used as the back gate electrode. Ti/Au metal layers were defined using electron-beam lithography, and the single ultralong ZnO NW was immobilized by silver paste carefully. After that, the electrodes were passivated with polymethyl methacrylate (PMMA) to reduce the leak current and eliminate the effect of metal-nanowire contact region to make sure the entire conductance changes originate from the nanowire [18].

2.3. *Surface modification of ZnO nanowire*

In order to enhance the immobilization efficiency, we adopted the widely used cross-linking method rather than the physical absorption via the following steps. The ZnO NWs were treated with an oxygen plasma (0.3Torr, 25W power for 60 s) to remove contaminants and add hydroxyl groups to the surfaces, and were then immersed in a 2% ethanol solution of 3-aminopropyltriethoxysilane (APTES). After the reaction, the amino-silanized surface was rinsed with ethanol thrice and then was baked at 120 °C for 10 min under N_2 gas. After wire bonding, 5μl uricase (5 units/ml) was deposited on the surface of ZnO nanowire, and the bioFET device was placed in saturated glutaraldehyde vapor for 40 minutes. The device was then rinsed with 0.01×PBS and deionized water for 15 minutes before air-dried. After modification, the device was covered with the solution exchanging chamber, and kept in 0.01 × PBS at 4°C before further calibration and measurements.

2.4. *Measurement of real-time conductance change of ZnO NW bioFET*

The conductance change of the ZnO NW bioFET functionalized with uricase was measured with semiconductor parameter analyzer (Keithley 4200) for exposure to uric acid with the concentration tenfold increased every time from 1pM to 0.5mM in a 0.01×PBS, and the V_g (gate voltage) and V_{sd} (source-drain voltage) were fixed at 0V and -1V, respectively. All the electrical measurements were performed at room temperature in an ambient air environment.

3. Results and discussion

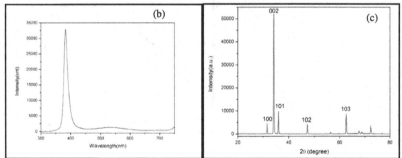

Fig. 1. (a) Top-view SEM image of ZnO nanowires grown on silicon substrate; (b) XRD pattern of ZnO nanowires and (c) photoluminescence spectrum of ZnO nanowires.

Fig. 1 (a) shows the typical FE-SEM (FE-SEM, LEO1530, Japan) image of as-synthesized ZnO nanowires. The average diameter is about 500nm, and the length could be longer than 1mm. We have also assessed the crystal structure and phase purity of the bulk nanowire samples using XRD (XRD, D/MAX-RB), Fig. 1 (b) as below. All the relatively sharp diffraction peaks are in good agreement with the standard ZnO wurtzite structure. The strongest peak (002) means that the growth of zinc oxide has a very clear C-axis orientation. Fig. 1(c) shows the room temperature photoluminescence spectrum of the NWs measured by using a continuous He-Cd laser (325 nm) as an excitation source. The extremely strong UV emissions at ~380 nm and negligible deep-level emissions at ~520 nm imply that the NWs have defect-suppressed crystal structures.

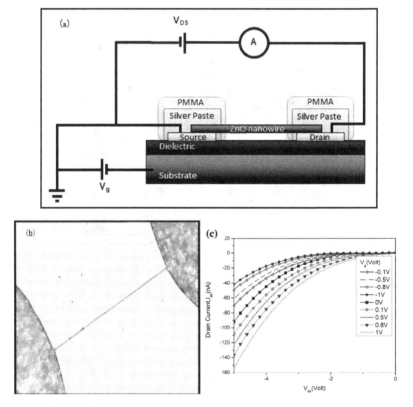

Fig. 2. (a) Schematic of single ultralong ZnO nanowire biosensor system and (b)Optical image of biosensor (c) I_{ds}-V_{ds} measurements under varying V_g, V_{ds}=-1V.

Using the as-synthesized high quality NWs, the ZnO NW FETs were fabricated using standard procedures with back gate geometry .The Fig. 2 (a) illustrates the schematic of ZnO nanowire biosensor system and (b) shows the optical image of it. We have then measured the device characteristics as shown in Fig. 2 (c), and the drain current (I_{ds}) versus drain voltage (V_{ds}) characteristics of ZnO NWs-based transistors obtained as a function of different gate voltages (V_g) indicate that the pronounced gate effect is indicative of an n-type semiconductor. The on/off ratio and transconductance were ~4.6×10^6 and ~8.2 nS, respectively.

Prior to the electrical measurement of the ZnO NW bioFET, optical images were obtained in order to investigate the modification of the ZnO NW surface with uricase. Under the high magnification optical microscope (×2000), the Fig. 3 (c) shows the ZnO NW has a transparent layer on the surface. It indicates that the functionalization methods were available for the modification of ZnO NW surface. The uricase immobilized ZnO NW permitted the binding of uric acid, which could affect the electrical characteristics of ZnO NW biosensor.

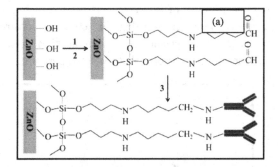

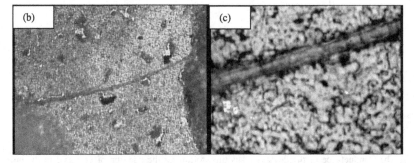

Fig. 3. (a) Immobilization of Uricase onto the ZnO NW surface via crosslinking surface modification method (i)2% 3-APTES in ethanol; (ii) 25wt% GAD; (iii) 5µl uricase (5 units/ml) (b) Optical image of ZnO nanowire crosslinking surface modification at low magnification ; (c) high magnification image

The fabricated NW FET was loaded into the home-made liquid exchanging chamber with the sensor electrically wired to the semiconductor parameter analyzer. After the baseline signal was established in pure 0.01×PBS buffer, aliquots of a 10 mg/mL solution of BSA were added for two main reasons: First, BSA can efficiently block the unmodified site to reduce the nonspecific binding interactions which may lead to false positive results and hence decrease the signal to noise ratio; Second, the increase of the protein concentration in the buffer could help to keep the enzymatic activity of uricase. After each BSA addition [2, 8, 12], the baseline re-equilibrated to a lower value, as shown in Fig. 4 (a). When the conductance baseline was stable in the protein-rich medium, we added the uric acid to the solution from 1pM to 0.5 mM with the concentration tenfold increased every time in 0.01×PBS, as Fig. 4 (b) below. The conductance of the device rapidly increased upon exposing the nanowire sensor to uric acid. Under our measurement conditions, the response time turned out to be in the order of millisecond which can be considered relatively short compared to other diagnostic technologies. The calibrated relationship between the saturated current and the concentration of uric acid was plotted in Fig. 4 (c). From the plot, we can come to the conclusion that the sensor as constructed was more sensitive at low uric acid concentrations than at high concentrations. All electrical measurements were performed at room temperature.

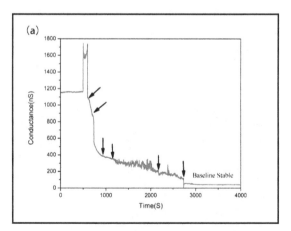

Fig. 4. (a) The Conductance-time curve of BSA passivation process, the arrow means the addition of BSA; (b) Conductance of the device versus time following the addition of uric acid in the buffer solution. The upper inset picture is the homemade reaction cell for sensing, and the optical image of silver paste immobilized ultralong ZnO nanowire; the lower inset is the schematic of our device during active sensing measurements. (c) Plot of saturated current vs. different concentrations of uric acid.

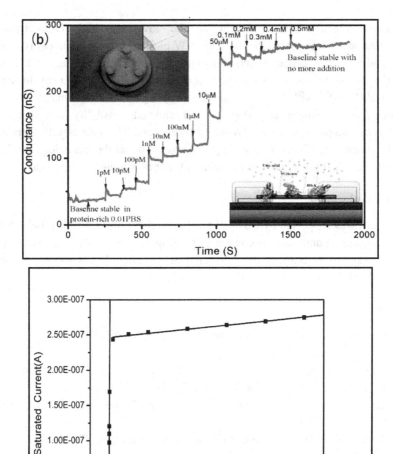

Fig. 4. (Continued)

4. Conclusion

ZnO NWs were directly synthesized through the CVD method. The characterization results of XRD and PL spectrum implied that the NWs had a well-crystallized structural quality. Using the as-synthesized high quality nanowire, electrical biosensors based on FET were fabricated for the highly sensitive detection of uric acid at the low and high concentrations simultaneously. It is shown that the ZnO NW bioFET sensors could easily detect uric acid down to a concentration of 1 pM with conductance increase of 14.7 nS, and the response time turns out to be in the order of millisecond. In conclusion, a single ZnO NW with an actual n-type property was easily fabricated as a nanobiosensor without any doping, and showed feasibility and a higher sensitivity compared to other NW-based biosensors. This cost effective process can be further exploited by expanding into arrays so that the technology can lead to portable, reliable and real-time detection for applications in many areas.

Acknowledgement

This work was supported by NSFC (51172022, 50972011), NSAF (10876001), the Research Fund of Co-construction Program from Beijing Municipal Commission of Education, the Fundamental Research Funds for the Central Universities and the Beijing novel program (2008B19).

References

1. X. H.Wang, Y.Chen, K. A.Gibney, S.Erramilli, P.Mohanty, Appl. Phys. Lett. **92**, 013903(2008).
2. C.Li,M.Curreli,H.Lin,B.Lei,F.N.Ishikawa,R.Datar,R.J.Cote,M.E.Thompson ,C.W.Zhou,J.Am.Chem.Soc.**127**(36),12484-12485(2005).
3. J.Liu, J.Goud, P. M.Raj, M.Iyer, Z.Wang, R. R.Tummala, Elec Comp C. 1971-1976 (2007).
4. V.Vamvakaki, K.Tsagaraki, N.Chaniotakis, Anal. Chem. **78**(15), 5538-5542 (2006).
5. T.Cohen-Karni, Q.Qing, Q.Li, Y.Fang, C. M.Lieber, Nano.Lett.**10** (3), 1098-102(2010).
6. G.Zheng, F.Patolsky, Y.Cui, W. U.Wang, C. M.Lieber, Nature biotechnology. **23**(10), 1294-301(2005).
7. F.Patolsky, G.Zheng, O.Hayden, M.Lakadamyali, X.Zhuang, C. M.Lieber, P.N.A.S. **101**(39), 14017-22(2004).
8. Y.Cui, Q.Wei, H.Park, C. M.Lieber, Science. **293** (5533), 1289-92(2001).
9. J.I.Hahm, C.M.Lieber, Nano.Lett.**4**(1), 51-54(2003).
10. W. U.Wang, C.Chen, K. H.Lin, Y.Fang, C. M.Lieber, P.N.A.S. **102** (9), 3208-12(2005).
11. C. M. Lieber, Mrs.Bull. **28**(2003).
12. F.Patolsky, C.M.Lieber, Materials.Today. **8**(4), 20-28(2005).
13. J.Zhou, N. S.Xu, Z. L.Wang, Adv. Mater.,**18** (18), 2432-2435(2006).

14. F.Patolsky, B. P.Timko, G.Yu, Y.Fang, A. B.Greytak, G.Zheng, C. M.Lieber,Science.**313**(5790),1100-4(2006).

15. M.Hernandez-Velez, Thin Solid Films. **495**(1-2), 51-63(2006).

16. F.F.Zhang, X. W, S.Y.Ai,Z.D.Sun,W. Qiao,Z.Q. Zhu,Y.Z. Xian,L.T. Jin,K.Yamamoto,Anal.Chem.Acta. **519**(2), 155-160(2004).

17. S. P.Singh, S. K.Arya, P.Pandey, B. D.Malhotra, S.Saha, K.Sreenivas, V.Gupta, Appl.Phys.Lett. **91**, 063901(2007).

18. R. J.Chen, H. C.Choi, S.Bangsaruntip, E.Yenilmez, X.Tang, Q.Wang, Y. L.Chang, H.Dai, J. Am. Chem. Soc. **126** (5), 1563-8(2004).

A HIGH-PERFORMANCE GLUCOSE BIOSENSOR BASED ON ZNO NANOROD ARRAYS MODIFIED WITH AU NANOPARTICLES

GONG ZHANG, YANG LEI, XIAOQIN YAN[†]

State Key Laboratory for Advanced Metals and Materials, School of Materials Science and Engineering, University of Science and Technology Beijing, Beijing 100083, People's Republic of China
and Key Laboratory of New Energy Materials and Technologies, University of Science and Technology Beijing, Beijing 100083, People's Republic of China

An amperometric glucose biosensor based on vertically aligned ZnO nanorod (NR) arrays modified with Au nanoparticles (NPs) was constructed in a channel-limited way. Au NPs with diameters in the range of 8-10 nm have been successfully synthesized by photo-reduction method and were uniformly loaded onto the surface of ZnO NRs that was hydrothermally deposited on the Fluorine doped SnO_2 conductive glass (FTO) via electrostatic self-assembly technique. The morphology and structure of Au/ZnO NR arrays were characterized by field-emission scanning electron microscopy (FE-SEM), high-resolution transmission electron microscopy (HRTEM) and X-ray photoelectron spectrum analyzer (XPS). The electrocatalytic properties of glucose oxidase (GOD)-immobilized Au/ZnO NR arrays were evaluated by amperometry. Compared with the biosensor based on ZnO NR arrays, the resulting Au/ZnO NR arrays modified biosensor exhibited an expanded linear range from 3 µM to 3 mM with the detection limit of 30 nM and a smaller Michaelis-Menten constant of 0.7836 mM. All these results suggest that the Au NPs can greatly improve the biosensing properties of ZnO NR arrays and therefore Au/ZnO NR arrays provide a promising material for the biosensor designs and other biological applications.

1. Introduction

In the past decade, the investigation of the application of hybrid nanomaterials in biosensing has been one of the most extensively studied research areas for their combination of the different properties of each component [1-3]. Nanomaterials provide high surface areas for higher enzyme loading and a compatible microenvironment helping the enzyme to retain its bioactivity. Besides, they provide direct electron transfer between the enzyme's active site and the electrode [4-7]. Among the nanomaterials, ZnO with a wide band gap

---

[†] Corresponding author, e-mail: xqyan@mater.ustb.edu.cn

has attracted much attention because of its favorable properties including biocompatibility, low toxicity, high electron mobility and easy fabrication [8, 9]. Furthermore, ZnO NMs with high isoelectric point (pI~9.5) are suitable for the adsorption of proteins or enzymes with low pI (e.g. glucose oxidase (GOD), pI=4.2-4.5) at physiological pH of 7.4 by electrostatic attraction [10, 11]. These capabilities make ZnO a favorable material for immobilization of biomolecules without an electron mediator and it can be employed for developing implantable biosensors [12-15].

Attempts to enhance the properties of ZnO have been made by doping or modifying with transition metals or noble metals [16-18]. As is known, Au is one of the only two transition metals more electronegative than Pt, and therefore it is expected to improve the performance of ZnO by taking advantage of its biocompatibility, huge surfaces and good electro-catalytic activity. On top of many successful demonstrations, it remains a great challenge to prepare large areas of one-dimensional ZnO nanostructure especially for ZnO NR arrays modified with Au NPs, together with well-controlled dimensions and morphologies.

Our previous work has showed that an enhanced sensitivity of the grown ZnO biosensor was found to be much higher than that of the transferred ZnO biosensor [19]. In this paper, we have explored a simple, efficient and economical route to synthesize Au/ZnO NR arrays on the FTO glass through photo-reduction method [20-27] together with electrostatic self-assembly technique [28]. And a new working enzymatic glucose biosensor has been made based on Au/ZnO NR arrays in a way of constructing micro-fluidic channel which is favorable for real-time detection, as seen in Figure 1. Glucose oxidase (GOD) was immobilized on the Au/ZnO NR arrays by cross-linking method [29-32], and the glucose sensing performance of the as-prepared bioelectrode has been characterized. Compared with the biosensor based on ZnO NR arrays, the biosensor using Au/ZnO NR arrays showed a wider linear range, a lower detection limit and higher sensitivity. The results demonstrate that the Au/ZnO NR arrays can offer a new and promising immobilization material for the biosensor designs.

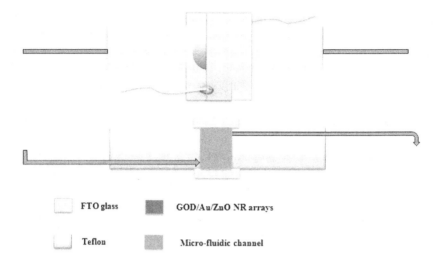

| | FTO glass | | GOD/Au/ZnO NR arrays |
| | Teflon | | Micro-fluidic channel |

Figure 1. Schematic diagram of Au/ZnO NR arrays biosensor

2. Experimental

2.1. Materials

Glucose (99%) and glutaraldehyde (50% solution) were purchased from Beijing Chemical Reagent Company. Nafion solution (5 wt. %), Glucose oxidase (GOD, 113U/mg) and bovine serum albumin (BSA>98%) were obtained from Sigma. All the reagents were analytical grade and used without further purification. 0.01M phosphate buffer solution (PBS) was prepared from NaH_2PO_4 and Na_2HPO_4, the pH was adjusted to 7.4 by NaOH.

2.2. Synthesis of Au/ZnO NR arrays

Aligned ZnO NR arrays on FTO glass were synthesized via hydrothermal approach. First, a layer of ZnO NPs was formed on the surface of FTO glass by thermally decomposing of zinc acetate at 300℃ for 30 min. Then the treated FTO glass was suspended in an aqueous solution containing 0.05M zinc nitrate and hexamethylenetetramine at 90℃ for 16 h. On completion of the reaction, the FTO glass was taken out from the solution and rinsed with deionized water. The resulting substrate was dried in air.

The typical procedure for depositing Au NPs on the ZnO NR arrays is given as follows. 1% HAuCl4 aqueous solution was diluted as 0.05% reaction solution with 1% PVA added as dispersant. In the photo-reduction of Au onto the ZnO, methanol was added as a sacrificial donor (at a molar ratio methanol: noble metal salts=500: 1). The pH of the water solution was adjusted by adding NaOH until reaching the pH value of 7. Then, FTO glass with ZnO NR arrays was then immersed in the reaction solution. A 300 W UV-C lamp (OSRAM, 300 W) hung over the photo-deposition reactor was used as the light source. After irradiated for 15 min, FTO glass with as-obtained ZnO NR arrays containing photo-reduced Au was washed with deionized water and then annealed in air at 300°C for 3 h to remove the capped PVA. Thus, the ZnO NR arrays with various Au loadings were prepared.

2.3. *Glucose biosensor fabrication*

Before the immobilization of GOD, ZnO NR arrays modified electrodes was rinsed with PBS to generate a hydrophilic surface. An enzyme solution was prepared by dissolving 10 mg GOD and 20 mg BSA in 200 µl PB solution. 2 µl of the above mixture was applied onto the surface of as-prepared electrode. The electrode was then left in air to dry which also allowed GOD and BSA to adsorb onto the ZnO NR arrays. The above procedure was repeated. The cross-linking procedure was carried out by adding 1.5 µl aqueous solution containing 2.5% glutaradehyde and 0.5% Nafion onto the electrode for two times. After dried at room temperature, 1.5 µl of 0.5% Nafion solution was further dropped onto the electrode surface to prevent possible enzyme leakage and eliminate foreign interferences. Finally, the electrode was immersed in PB solution to remove unimmobilized enzymes. All prepared enzyme electrodes were stored in dry condition at 4°C when not in use. For the comparison of the significance of the modification of Au NPs, two electrodes based on the ZnO NR arrays and Au/ZnO NR arrays were made with the identical procedure, respectively.

2.4. *Biosensor characterization and electrochemical measurements*

The morphology and structure of ZnO NR arrays and Au/ZnO NR arrays were characterized by field emission scanning electron microscopy (FE-SEM, Zeiss, SUPRA-55, Germany), high-resolution transmission electron microscopy (HRTEM), selected area electron diffraction (SAED), energy dispersive spectrometer (EDS) and X ray photoelectron spectrum analyzer (XPS). The cyclic voltammograms (CV) were acquired from -0.2 V to +0.8V at a scan rate of 20 mV/s in 0.01M PB solution of pH 7.4. The amperometric response of biosensor to glucose was recorded at +0.8V. All electrochemical experiments were performed at room temperature (typically, 25°C).

3. Results and discussion

To avoid the shielding of active sites of ZnO and the poisoning of catalytic activity of Au/ZnO by functional organics, the electrostatic self-assembly [28] was employed to directly adsorb charged Au NPs on unmodified ZnO NR arrays. The base principle can be described as follows: By adjusting the pH of reaction solution to the value of pH 7, Au NPs protected by PVA are negatively charged while ZnO NR arrays are positively charged. Thus Au NPs with PVA will spontaneously adsorbed on ZnO NR arrays through electrostatic attraction, as is depicted in Figure 2. Here PVA which can be easily removed by sintering also functions as a dispersant preventing coalescence from Au NPs.

Figure 3 (a) shows that vertically aligned ZnO NR arrays coated by Au NPs were synthesized on the FTO glass. The average diameter of ZnO NRs varies from ~50 nm to ~100 nm. The distribution of the Au NPs with an average diameter in the range of 8~10 nm over individual ZnO NR is fairly uniform as depicted in the TEM image in Figure 3 (b). The small size of Au NPs shows the good thermal stability of loaded Au NPs in the calcinations. It can be also seen from the picture that the length of the ZnO NRs is about 1 μm. To further reveal the micro-stucture of the Au-ZnO composite, HRTEM image and corresponding SAED have been recorded as shown in Figure 3 (c). The interplanar distance of fringes of ZnO NR was measured to be 0.26 nm. ZnO NRs grew uniformly along the direction of (0001) plane with no obvious defects. The loaded Au NPs, with an approximate spherical shape, have the diameter of 8~10 nm, which is well agreed with the TEM observation.

EDS pattern from Figure 4 (a) indicates that the above Au-ZnO hybrid nanostructure is composed of the elements of Zn, O, Sn and Au, which should be ascribed to ZnO NR arrays, FTO glass and Au NPs. In order to exactly investigate the valence state of Au in the resulting sample, an XPS experiment has been used for the analysis of Au NPs. Figure 4 (b) displays the XPS spectrum of the Au 4f region. Au peaks are located at 84.1 and 87.8 eV corresponding to the electronic states of Au $4f_{7/2}$ and Au $4f_{5/2}$, respectively, which are highly consistent with the literature [33]. Results from the spectral area integration of individual components indicate that more than 90% of the Au exists as zero valence species, whereas the other represents species in oxidation states. The high percentage of metallic species suggests a highly effective reduction of reaction solution via photo-reduction method.

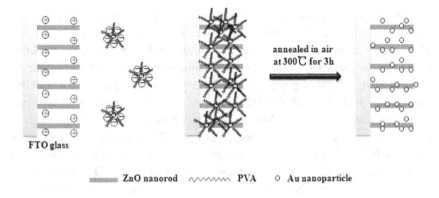

Figure 2. Schematic diagram of ZnO NRs array modified with Au NPs

Figure 3. (a) FESEM image of Au/ZnO NRs array, (b) TEM image and (c) HRTEM image of single Au/ZnO NR (the inset is the SAED pattern)

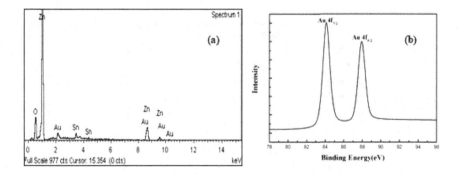

Figure 4. (a) XPS pattern and (b) EDS pattern of Au/ZnO NR arrays

The enzymatic generation of GOD was achieved in the reaction layer of the Au/ZnO NR arrays. To investigate the electro-catalytic behavior toward the electrochemical reaction of GOD at the Au/ZnO NR arrays film, the modified electrode was characterized by a cyclic voltammetric (CV) sweep curve ranging from -0.2V to +0.8V at a scan rate of 20 mV/s. For comparison, a controlled experiment with GOD/ZnO/FTO electrode was performed to record the response as seen in Figure 5. Well-defined redox peaks can be observed at both of the above two electrodes. For the GOD/ZnO/FTO electrode, peak difference (ΔEp) is 0.334 V, indicating a quasi-reversible redox process. An observable ΔEp estimated as 0.160 V was obtained at the GOD/Au/ZnO/FTO electrode, which could be ascribed to the increased electron transfer ability and electrocatalytic ability introduced by Au NPs.

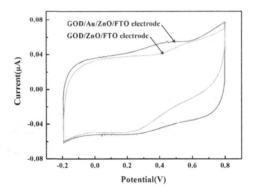

Figure 5. CVs obtained at the GOD/ZnO/FTO electrode (green line) and the GOD/Au/ZnO/FTO electrode (red line) in 0.01M PB solution at pH 7.4

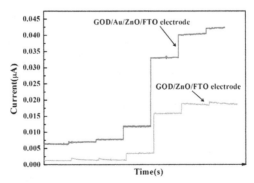

Figure 6. Amperometric response of the biosensors based on ZnO NR arrays and Au/ZnO NR arrays to different glucose concentration at +0.8V

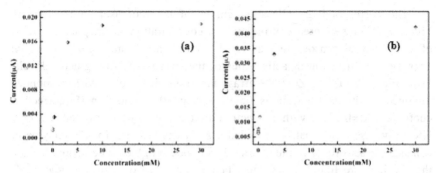

Figure 7. Calibrated curve of the biosensor with successive increase of glucose concentration based on (a) ZnO NR arrays and (b) Au/ZnO NR arrays

With respect to the benefit of the Au NPs in the construction of biosensors, comparison was performed by measuring the amperometric response of biosensors using ZnO NR arrays and Au/ZnO NR arrays. Figure 6 shows the amperometric response of the ZnO NR arrays modified biosensor and the Au/ZnO NR arrays modified biosensor to the different glucose concentration solutions at an applied potential of +0.8V. On one hand, it has been revealed that both of the biosensors exhibited a rapid and sensitive response to the change of glucose concentration. Well-defined steady-state current responses were obtained and the currents increased stepwise with successive raise of glucose concentration, which indicate a good electro-catalytic oxidative and fast electron exchange behavior of the modified electrodes. On the other hand, the difference between the amperometric responses of the two modified electrodes can be clearly observed in the figure. The response currents of the Au/ZnO NR arrays modified biosensor are much higher than that of the ZnO NR arrays modified biosensor when detecting the same glucose concentration. This demonstrates that ZnO NR arrays modified biosensor loaded with Au NPs has been a substantial improvement to the performance of the original biosensor due to the high specific surface area, high activity, high catalytic efficiency and good biocompatibility of Au NPs. The system studied in this work can be depicted as follows:

$$Glucose+GOD/FAD \rightarrow Gluconic\ acid+GOD/FADH_2 \qquad (1)$$

$$GOD/FADH_2 \rightarrow GOD/FAD+2H^++2e^- \qquad (2)$$

The corresponding calibration curves of the biosensors are shown in Figure 7. With the increase of the glucose concentration, the response currents of both two biosensors increase and saturate at 10 mM glucose, which suggests the saturation of active sites of the enzyme at those glucose levels. From Figure 7 (a), the ZnO NR arrays modified biosensor linearly responded to glucose with concentrations ranging from 30 μM to 3 mM (correction coefficient R^2=0.9971) with a detection limit of 3 μM, which proved that ZnO NR arrays used as a matrix provided a good environment for the enzyme to enhance the sensitivity of the modified electrode for glucose detection. In fact, these values are found to be much better that those of many of other ZnO nanomaterials-based biosensors reported previously [12, 15, 34, 35].The sensitivity of the ZnO NR arrays modified biosensor is calculated to be 0.0192 μA·cm^{-2}·mM^{-1}.The apparent Michaelis-Menten constant ($K_{app,M}$), which is a measure of the enzyme-substrate kinetics of the glucose biosensor, can be calculated from the Lineweaver-Burk equation [36]:

$$\frac{1}{I} = \frac{K_{app,M}}{I_{max}}\frac{1}{C} + \frac{1}{I_{max}} \tag{3}$$

where C is the glucose concentration, Imax and I are the currents measured for enzymatic product detection under conditions of substrate saturation and steady state, respectively. A plot if $1/I$ vs. $1/C$ will give a straight line with the slope equal to $K_{app,M}/I_{max}$, and intercept equal to $1/I_{max}$. The $K_{app,M}$ value calculated from the Lineweaver-Burk equation for the GOD/ZnO/FTO bioelectrode was 1.5843 mM. The small $K_{app,M}$ means that the immobilized GOD possessed a high enzymatic activity, and the proposed modified electrode exhibited a high affinity for glucose [37]. In contrast with ZnO NR arrays biosensor, the calibration plot of Au/ZnO NR arrays biosensor for glucose detection in Figure 7 (b) shows a wider linear response from 3 μM to 3 mM with a correlation coefficient of 0.990. The detection limit was determined to be as low as 30 nM. Besides, the sensitivity and $K_{app,M}$ value were calculated to be 0.0336 μA·cm^{-2}·mM^{-1} and 0.7836 mM (R^2=0.9995), respectively. From the comparison data mentioned above, it is easy to find that Au/ZnO NR arrays biosensor exhibited much more excellent performance than ZnO NR arrays biosensor in the aspects of detection limit, sensitivity and linear range. Therefore, we can conclude that the as-synthesized Au-decorated ZnO NR arrays have better electro-catalytic activity and electron transfer ability towards glucose when compared with ZnO NR arrays alone.

4. Conclusions

In summary, we have successfully synthesized Au/ZnO NR arrays via photo-reduction method combined with electrostatic self-assembly technique. Au NPs were uniformly modified onto the ZnO NR arrays with the mean diameter of around 8~10 nm. It was found that both of the ZnO NR arrays biosensor and the Au/ZnO NR arrays biosensor exhibited wide linear range, low detection limit, short response time and high affinity for glucose. However, with Au NPs modified onto the ZnO NR arrays, the biosensor showed superior electro-catalytic activity and electron transfer ability towards glucose. The Au/ZnO NR arrays biosensor displayed a relatively wide linear range of 3 μM~3 mM, a low detection limit of 30 nM and a relatively small Michaelis-Menten constant of 0.7836 mM. Importantly, to the best of our knowledge, this is the first time that a glucose biosensor with these favorable properties has been achieved by using a GOD/Au/ZnO/FTO modified electrode. Hence, Au/ZnO NR arrays provide a new platform for biosensor design and other biological applications.

Acknowledgments

This work was supported by NSFC (51172022, 50972011), NSAF (10876001), the Research Fund of Co-construction Program from Beijing Municipal Commission of Education, the Fundamental Research Funds for the Central Universities and the Beijing novel program (2008B19).

References

1. X. H. Wang, Y. Chen and K. A. Gibney, Appl. Phys. Lett. **92**, 013903 (2008).
2. B. X. Gu, C. X. Xu and G. P. Zhu, J. Phys. Chem. B **113**, 6553 (2009).
3. Y. H. Lin, F. Lu, Y. Tu and Z. F. Ren, Nano Lett. **4**, 2 (2004).
4. C. Jagdish and S. J. Pearton, Thin Films and Nanostructures; Elsevier: Oxford, U. K. p 600 (2006).
5. Q. Xua, C. Mao, N. N. Liu, J. J. Zhu and J. Sheng, Biosens. Bioelectron. **22**, 768 (2006).
6. Y. Xiao, F. Patolsky, E. Katz, J. F. Hainfeld and I. Willner, Science **299**, 1877 (2003).
7. J. Jia, B. Wang, A. Wu, G. Cheng, Z. Li and S. A. Dong, Anal. Chem. **74**, 2217 (2002).
8. Z. R. R. Tian, J. A. Voigt and J. Liu, J. Am. Chem. Soc. **124**, 12955 (2002).
9. L. Vayssieres, Adv. Mater. **15**, 5 (2003).

10. Y. Lei, X. Q. Yan, N. Luo, Y. Song and Zhang Y., Coll. Surf. A: Physicochem. Eng. Aspe. **361**, 169 (2010).

11. D. Pradhan, F. Niroui and K. T. Leung, Appl. Mater. Inter. **8**, 2409 (2010).

12. A. Wei, X. W. Sun, J. X. Wang, Y. Lei, X. P. Cai, C. M. Li, Z. L. Dong and W. Huang, Appl. Phys. Lett. **89**, 123902 (2006).

13. S. Kumar, V. Gupta and K. Sreenivas, Nanotechnology **16**, 1167 (2006).

14. S. Krishnamurthy, T. Bei, E. Zoumakis, G. P. Chrousos and A. Iliadis, Biosens. Bioelectron. **22**,707 (2006).

15. J. X. Wang, X. W. Sun, A. Wei, Y. Lei, X. P. Cai, C. M. Li and Z. L. Dong, Appl. Phys. Lett. **88**, 233106 (2006).

16. A. K. M. Kafi, G. S. Wu and A. C. Chen, Biosens. Bioelectron. **24**, 566 (2008).

17. D. Wen, S. J. Guo, Y. Z. Wang and S. J. Dong, Langmuir, **26**, 11401 (2010).

18. J. E. Gregory, R. Aimee, Q. W. Li and Jr. F. W. Charles, J. Vac. Sci. Technol. A **16**, 1926 (1998).

19. Y. Lei, X. Q. Yan, J. Zhao, X. Liu, Y. Song, N. Luo and Y. Zhang, Coll. Surf. A: Physicochem. Eng. Aspe. **361**, 169 (2010).

20. A. Sclafani and J. M.Herrmann, J. Photochem. Photobiol. A: Chem. **113**, 181 (1998).

21. T. Sano, S. Kutsuna, N. Negishi and K. Takeuchi, J. Mol. Catal. A: Chem. **189**, 263 (2002).

22. M. I. Litter, Appl. Catal. B **23**, 89 (1999).

23. D. Hufschmidt, D. Bahnemann, J. J. Testa, C. A. Emilio and M. I. Litter, J. Photochem. Photobiol. A: Chem. **148**, 223 (2002).

24. C. Crittenden, J. Liu, D. W. Hand and D. L. Perram, Water Res. **31**, 429 (1997).

25. B. Kraeutler and A. J. Bard, J. Am. Chem. Soc. **100**, 4317 (1978).

26. S. Zheng and L. Gao, Mater. Chem. Phys. **78**, 512 (2002).

27. V. Iliev, D. Tomova, L. Bilyarska and L. Petrov, Catal. Commun. **5**, 759 (2004).

28. P. F. Fu and P. Y. Zhang, App. Catal. B: Environmental **96**, 176 (2010).

29. T. Kong, Y. Chen, Y. P. Ye, K. Zhang, Z. X. Wang and X. P. Wang, Sens. Actuat. B: Chem. **138**, 344 (2009).

30. H. Muguruma, A. Hiratsuka and I. Karube, Anal. Chem. **72**, 2671 (2000).

31. Q. Yang, P. Atanasov and E. Wilkins, Sens. Actuat. B: Chem. **46**, 249 (1998).

32. B. Wu, G. Zhang, S. Shuang and F. M. M. Choi, Talanta **64**, 546 (2004).

33. J. F. Moulder, W. F. Stickle, P. E. Sobol and K. D. Bomben, Handbook of X-ray photoelectron Spectroscopy. Physical Electronics Inc. Eden Prairie MN. (1995).
34. D. Pradhan, F. Niroui and K. T. Leung, App. Mater. Inter. **2**, 2409 (2010).
35. Z. W. Zhao, X. J. Chen, B. K. Tay, J. S. Chen, Z. J. Han and K. A. Khor, Biosens. Bioelectron. **23**, 135 (2007).
36. R. A. Kamin and G. S. Wilson, Anal. Chem. **52**, 1199 (1980).
37. Z. Wen and S. Ci, J. Phys. Chem. C **113**, 13482 (2009).

PIEZOTRONIC VIBRATION DETECTOR BASED ON ZNO NANOWIRE ARRAYS

ZHENG ZHANG, YUNHUA HUANG[†], QINGLIANG LIAO, PING LI,
SHENGPING CHEN

*Department of Materials Physics, State Key Laboratory for Advanced Metals and
Materials, University of Science and Technology Beijing
Beijing, 100083, China*

Due to the semiconducting and piezoelectric properties, ZnO nanowires have demonstrated novel applications on vibration detecting. In this study, a vibration detector based on ZnO nanowire arrays (ZnO-NWAs) was constructed, which was employed to detect the frequency of the electromagnetic vibrator. The results show that the Schottky contacts appeared between ZnO-NWAs and Pt electrodes, and the current of the device increases with the compressive strain on the flexible electrodes. The strain response of the detector was suggested to be attributed to both piezoelectric and piezoresistance effects of ZnO-NWAs. These devices constructed entirely were found to have a high sensitivity (above 5000%) and a short response time to the frequency of compressive strain (about 0.2-15Hz).

1. Introduction

Vibration is the most common phenomenon that exists not only in our daily lives, but also in the microscopic world. Sometimes vibration does harm to the connection unit of the machinery equipment support and even the structure of critical infrastructures under extreme conditions. Therefore how to detect and utilize vibration have attracted much more attention. Recently, micro and nano electro-mechanical system (MEMS and NEMS) is concerned much more with considerable potential for ultra-fast, high-sensitivity and low-power consumption devices. ZnO, as an important material that exhibits coupled semiconductor and piezoelectric properties, has a wide range of applications in novel devices fabrication. Recently, various novel nanodevices have been fabricated by using the coupled semiconductor and piezoelectric properties of ZnO nanowires, such as nanogenerators [1, 2], strain sensors [3-5], acoustic sensors [6], piezoelectric field effect transistors [7], and transducer and actuator [8]. However, there were few reports on vibration detector based on ZnO-NWAs. Wang's group elaborated the fundamental principle of nanopiezotronics

---

[†] Corresponding author. *E-mail address:* huangyh@mater.ustb.edu.cn.

[9], which utilized the coupled piezoelectric and semiconducting properties of nanowires and nanobelts for designing and fabricating electronic devices. They also suggested that the charge carrier transport process in FET could be tuned by applying a stress to the device which could change the piezopotential in a piezoelectric nanowire. This type of device could be driven or triggered by a mechanical deformation action.

2. Experimental

In this letter, we reported a new type of vibration detector that is built using vertically grown ZnO-NWAs. Its operation mechanism relies on the periodic change of the conductance in ZnO-NWAs with cyclic press due to both the piezoresistance effect and piezoelectric effect.

2.1. Preparation of the ZnO nanowire arrays

The ZnO-NWAs used in this research were grown on fluorine-doped SnO_2 (FTO)-coated glass plates that were used as the substrates by low-temperature aqueous chemical growth method. Two processes were employed in this aqueous chemical growth method. One was sol-gel process to prepare the ZnO seed layer, and the other was hydrothermal process to grow ZnO-NWAs on the substrate. Firstly the FTO substrate cleaned by deionized water, ethanol, acetone, isopropanol in an ultrasonic bath each for 10min was blown under a stream of nitrogen and then baked at $150\,^{\circ}C$ for 15min. Colloid solution for the sol-gel process was prepared by dissolving zinc acetate $[Zn(CH_3COO)_2 \cdot 2H_2O]$ dried at $60\,^{\circ}C$ for 30min in ethylene glycol monomethyl ether (EGME) $[C_3H_8O_2]$ at room temperature. The molar concentration of zinc acetate was 0.5. Zinc nitrate hexahydrate $[Zn(NO_3)_2 \cdot 6H_2O]$ mixed with hexamethylenetetramine (HMTA) $[(CH_2)_6N_4]$ and Polyetherimide (PEI) was dissolved in deionized water to be used as the precursor solution in hydrothermal process. The molar concentration of $Zn(NO_3)_2$ and HMTA was fixed at 0.05mol/L and the concentration of PEI was 0.005mol/L. The colloid solution was coated onto the FTO substrate cleaned successfully by a spin coater at the rate of 5000 rad/min for 20s. The coated substrate was annealed at $350\,^{\circ}C$ for 30min to obtain dense and thin ZnO seed layer. Afterwards, the substrates coated by ZnO seed layer were placed in precursor solution at $95\,^{\circ}C$ for 12h without any stirring. Subsequently the substrate was taken out and cleaned smoothly.

2.2. Construction of vibration detector

The device was fabricated by growing vertical ZnO-NWAs on FTO substrate. A flexible polyimide (PI) film was cleaned with deionized water and ethanol under sonication, and coated with Pt about 200nm thick. Subsequently, the PI film coated with Pt was fixed on the top of the ZnO-NWAs. The Pt and FTO thin film were used as two electrodes of the detector. The Cu wires were used to connect the external circuitry. Finally, the device was packaged by PI paste. The schematic of the vibration detector was shown in Figure 1.

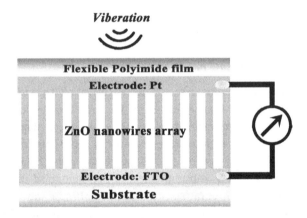

Figure 1. shows the schematic and the measurement setup of the vibration detector based on vertical growth ZnO-NWAs. The substrate is glass plate, and the electrodes are FTO film and PI film coated with Pt.

2.3. Measurement of the detector

The electromechanical properties of the ZnO-NWAs vibration detector were measured with semiconductor parameter analyzer (Keithley 4200) at room temperature in an ambient air environment. Figure 1 also shows the measurement setup. The substrate of detector was affixed on a sample holder, the flexible PI film was free to be pressed. A vibration cantilever beam with frequency resolution of 50Hz was used to press the free side of the detector to provide a regular frequency. The frequency and pressure to the detector could be regulated by a variable frequency generator.

3. Results and discussions

3.1. *Morphology and structures of ZnO-NWAs*

The morphology of ZnO nanowire array were examined by Field Emission Scanning Electron Microscopy (FE-SEM) and X-ray Diffraction (XRD). Figure 2 shows the surface morphology and crystal structure of the ZnO nanowire array. Figure 2a-b shows the top and side SEM images of ZnO-NWAs. The diameter of ZnO nanowires is in the range of 200~500nm, and the length is in the range of 4-5μm. Fig. 2c presents the XRD spectra of ZnO nanowire array showing the Bragg reflection corresponding to the typical ZnO wurtzite structure. Environmental fluctuations may have a negative effect on the number of species. This effect is due to physiological stress, which may even cause the extinction of some species.

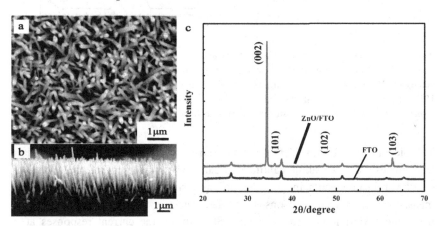

Figure 2. a-b shows the top and side SEM images of ZnO-NWAs. Figure 2c presents the XRD spectra of ZnO nanowire array showing the Bragg reflection corresponding to the typical ZnO wurtzite structure.

3.2. *Electrical properties of the detector*

Before the electromechanical measurement, we first tested the original I-V characteristic of the detector as shown in figure 3a. The nonlinear I-V characteristics were commonly observed in measuring the detector. Generally, the nonlinearity is caused by the Schottky barriers formed between the ZnO-NWAs and the Pt electrode in the detector, and the shape of the I-V curve depends on the heights of the Schottky barrier formed between the ZnO-NWAs and Pt electrode due to different interface properties [10, 11].

178

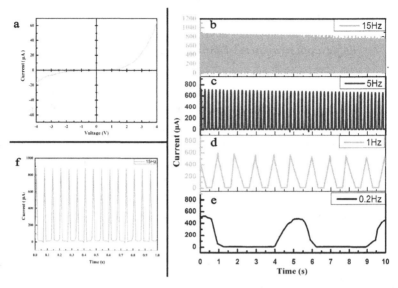

Figure 3a shows the I-V characteristic of the detector. The curve presents a typical Schottky barrier between the ZnO-NWAs and Pt electrode. Figure 3b-e shows the current responses at a fixed bias voltage of 3V under various periodic mechanical vibrations with frequencies of 0.2Hz, 1Hz, 5Hz and 15Hz, respectively. Figure3 f shows the current response under 15Hz in 1s. The curve shows that the detector has a high sensitivityand a short response time to the frequency of compressive strain.

The piezotronic vibration detector was used to detect the mechanical vibration frequencies of a vibration cantilever beam. The movement of the vibration cantilever beam results in periodic pressure strains on the ZnO-NWAs. We measured the current responses of the detector under different periodic mechanical vibration pressure. Figure 3b-e shows the current responses at a fixed bias voltage of 3V under various periodic pressures with frequency of 0.2Hz, 1Hz, 5Hz and 15Hz, respectively. It should be noticed that all of the data are the random parts of hundred times' experiments under every frequency, and short-time results are shown in figure 3f to display the current responses clearly.

It can be observed that the current increases with pressure, due to the coupled piezoelectric and piezoresistance effect. It has been reported that two typical effects are observed, with a single ZnO nanowire strained. The resistance change of the single ZnO nanowire was attributed to the change in band gap and possibly density of states in the conduction band. On the other, the piezopotential is created in the nanowire [12]. The ZnO-NWAs in the detector are viewed as lots of ZnO nanowires bonded on the FTO substrates vertically, as shown in figure 4. When the ZnO-NWAs were pressured, the schottky barrier height between ZnO nanowires and PT electrode is changed by the pizeopotential created along the nanowires and the change of band gap of each nanowire. Moreover, the band gap change also changes the intrinsic resistance of single ZnO nanowires due to the piezoresistance effect. However, the resistance of the nanowires array decreases obviously due to the increase of contact points between electrode and ZnO nanowires under the enhanced pressure.

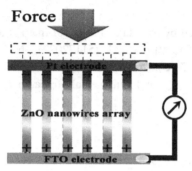

Figure 4. shows the schematic of the piezopotential created in ZnO-NWAs during the detector pressed.

For vibration detector, the sensitivity is an important parameter that can be calculated by:

$S = (I-I_0)/I$ (S is sensitivity, I is the response conductance, I_0 is the original conductance)

In this letter, the sensitivity of the detector to periodically mechanical vibration under different frequencies is above 5000% by calculating the first current peak of each data. The response time of the detector which was considered as another important parameter can be estimated to be about 12 ms from the shape of the current response under the frequency of 15Hz. The frequencies of the current responses shown in figure 3 b-e are calculated to be 0.2Hz, 1Hz, 5Hz, and 15Hz, respectively. And they are completely identical with the mechanical vibration frequencies in the software. The results indicate that the constructed vibration detector can be effectively used to detect the vibration frequency directly.

4. Conclusion

In summary, a piezotronic vibration detector is fabricated with vertically grown ZnO-NWAs on the FTO substrate and Pt coated flexible PI film. The I-V characteristics and the current responses under different frequencies demonstrate that the detector provides a good performance on detecting the frequency of

180

mechanical vibration directly. The current response under the vibration is attributed to coupling of the piezoresistance and piezoelectric effect. The detector fabricated is found to show a high sensitivity (above 5000%) and a fast response time (about 12 ms) to the vibration. These results support the applications of ZnO-NWAs in nanodevices.

Acknowledgments

This work was supported by the Major Project of International Cooperation and Exchanges (2006DFB51000), NSFC (51172022), NSAF (10876001), the Research Fund of Co-construction Program from Beijing Municipal Commission of Education, the Fundamental Research Funds for the Central Universities.

References

1. Z. L. Wang and J. Song, *Science*, **312**, 242 (2006).
2. S. N. Cha, J. Seo, S. M. Kim, H. J. Kim, Y. J. Park, S. W. Kim, and J. M. Kim, *Adv. Mater.*, **42**, 4726 (2010)
3. J. Zhou, Y. Gu, P. Fei, W. Mai, Y. Gao, R. Yang,G. Bao, and Z. L. Wang, , *Nano Letters*, **9**, 3035 (2008).
4. Y. Yang, J. J. Qi, Y. S. Gu, X. Q. Wang, and Y. Zhang, *Phys. Status solidi RRL*, **7-8**, 269 (2009).
5. H. Gullapalli, V. S. M. Vemuru, A. Kumar, A. Botello-Mendez, R. Vajtai, M. Terrones, S. Nagarajaiah, and P. M. Ajayan, *Small*,. **15**, 1641 (2010).
6. A. Arora, P. J. George, A. Arora, V. K. Dwivedi, and V. Gupta, *Sensors & Transducers*, **4**, 70 (2008).
7. X. D. Wang, J. Zhou, J. H. Song, J. Liu, N. S. Xu and Z.L. Wang, *Nano Lett.*, **6**, 2768 (2006).
8. B. A. Buchine, W. L. Hughes, F. L. Degertekin and Z. L. Wang, *Nano Lett.*, **6**, 1155 (2006).
9. Z. L. Wang, *Adv. Mater.*, **19**, 889 (2007).
10. Z. Y. Zhang, K. Yao, Y. Liu, C. H. Jin, X. L. Liang, Q. Chen, L. M. Peng, *AdV. Funct. Mater.*, **17**, 2478 (2007).
11. J. L. Freeouf, J. M. Woodall, *Appl. Phys. Lett.* **39**, 727 (1981).
12. Z. L. Wang, *J. Phys. Chem. Lett.*, **1**, 1388 (2010).

FABRICATION AND PROPERTIES OF A MICROSTRAIN SENSOR BASED ON ZINC OXIDE NETWORK STRUCTURE

PING LI, QINGLIANG LIAO[†], ZHENG ZHANG, SIWEI MA, YUE ZHANG

Department of Materials Physics, State Key Laboratory for Advanced Metals and Materials, University of Science and Technology Beijing
Beijing, 100083, China

ZnO materials are promising candidates for microelectronic mechanical systems for their excellent piezoelectric properties. In this study, a flexible strain sensor based on the unique ZnO micrometer-wire network (ZWN) structure was fabricated. The current characteristics of the device were investigated by applying tensile strains. This device also exhibits quick responses to tiny vibrations and microstrains with a high gauge factor up to 350. The effects of piezoelectric, piezoresistance and the ZWN structure were also analyzed in this work. The sensor has a wide range of applications in microrobots, electromechanical coupled electronics and acoustic detection.

1. Introduction

In recent years, the research of micro-electromechanical coupled electronics is a promising field for researchers all over the world. A number of approaches for nanoscale devices have been explored, such as solar cells [1-2], gas and liquid sensors [3-4] and biosensors [5]. These tiny devices enable human to explore the world and our body in a smaller scale. With monitoring of these ubiquitous and different signals, we could forecast the changes of environments or the work condition of the equipments and our body. To achieve the goal, materials with electromechanical properties are needed, such as $BaTiO_3$ [6], GaN [7] and Si [8].

ZnO is a material that exhibits both semiconductor and piezoelectric properties [9], it has a wide range of applications such as strain sensing [10], acoustic sensing [11], and energy harvesting [12], especially for flexible organic semiconductor technologies. All of the components of these sensors are built on flexible plastic films. Therefore, it is easy to place the electronic sheet over desks, floors, walls and any other location imaginable to detect tiny vibrations. However, microelectronic devices based on nanomaterials haven't realized mass

---

[†] Corresponding author. *E-mail address:* liao@ustb.edu.cn

production and commercial applications, and exact working mechanisms of these devices are not enough, which still needs more efforts in the future.

In this letter, a new type of flexible strain sensor based on ZnO micrometer-wire network (ZWN) structure with polyimide (PI) substrate was fabricated under low temperature in a simple way. Its performances are explored under both static strain and dynamic tiny vibrations with different frequencies. The results show that this sensor has good current responses and high sensitivity to tiny vibration detection.

2. Experimental Section

A hydrothermal route was employed to obtain ZnO micro materials. Firstly, the precursor solutions were prepared by dissolving $Zn(NO_3)_2 \cdot 6H_2O$, $(CH_2)_6N_4$ and Polyetherimide (PEI, low molecular weight) in distilled water. The equivalent concentrations of $Zn(NO_3)_2 \cdot 6H_2O$ and $(CH_2)_6N_4$ were 0.05M, respectively, and the content of PEI was 0.005M. Then, 200 ml of precursor solution was sealed in a glass bottle of maximum volume 400ml and then heated to 95°C for 3 h. At the end of the preheating process, the solution was filtered and sealed into another glass bottle, then heated to 95°C for 8 h to obtain ZnO materials.

PI film was ultrasonically cleaned in ethanol and acetone for 10 min, respectively. Then the ZnO microwires (0.2g) were soaked into ethanol and dispersed by ultrasonic wave for 30s to form a freely suspended ZnO micro wire solution. By dispersing a droplet of the solution onto the PI film, ZnO micro wires were fairly uniformly distributed on the substrate surface. Then the substrate was dried at a temperature of 30°C. To repeat this process for 3 times, we gained a equally distributed ZnO micro wire deposited layer on the PI substrate. For electrode fabrication, silver paste was applied at the two sides of ZnO layer. A thin layer of polydimethylsiloxane (PDMS) was used to package the device; providing enhancement of electrodes and resistance to corrosion. The entire device was annealing at 70°C for 3h. Finally, a flexible and well-packaged strain sensor device was fabricated, as figure 2.

The crystalline structure of the samples was characterized by using X-ray diffraction (XRD) system. The morphology of the ZnO layer was taken with a field-emission scanning electron microscope, as figure 1. The electric properties of the ZnO strain sensor were characterized by semiconductor characterization system (Keithley 4200-SCS).

3. Results and Discussion

Fig. 1 is the FESEM images of ZnO materials, indicating that well-dispersed ZnO micrometer wires with high density are observed on the surface of the PI substrate. The diameter of the micrometer wires are distributed in the range of 1μm with a length of 15μm. Each wire is well-connected to other wires serving as the electric tunnels in the entire device, which makes a ZWN structure between the two electrodes. Fig. 1(b) presents the XRD spectra of the ZnO layer, showing various Bragg reflections corresponding to the ZnO wurtzite structure. Figures 1(c) and (d) are optical images of PI film with a very thin ZnO layer, showing the flexibility of the PI substrate, which was fairly retained after coating with ZnO. Fig. 1(e) shows the picture of sealed strain sensor: Ag electrodes were firmly fixed on the two sides of PI film coated by PDMS thin membrane.

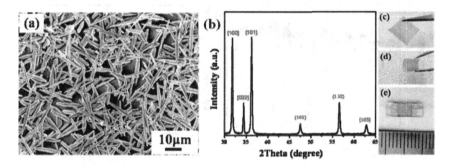

Fig. 1. (a) FESEM images of the well-dispersed ZWN on PI substrate coated by a thin PDMS layer; (b) XRD spectra of the ZWN structure; (c) and (d) are the optical images of PI film with a very thin ZnO layer, showing well flexibility of the substrate; (e) The picture of sealed strain sensor: Ag electrods on the two sides of PI film coated by PDMS.

The testing system for the strain sensor is shown schematically in Fig. 2(a), with one end fixed to a stand and the other end suspended with the strain loaded. Typical I-V characteristics under various static tensile strains are shown in Fig. 2(b). It is clear that the currents both under positive bias and negative bias are suppressed when the tensile strain is increased. Finally a downward I-V behavior is received. The response of this strain sensor to an external vibration can be estimated from the shape of the current curves as shown in Figures 2(c) and (d). The curve is sharply decreased when the substrate is slightly deformed by the vibrations. The frequency of the quick response corresponds well with the frequency of excitation with a input voltage of 1.5 V. The inset image of

184

Fig. 2(c) shows the details of a typical current response under the frequency of 0.2Hz. When the frequency was increased to 0.5Hz, evident current responses were still observed in Fig. 2(d). The results demonstrate that the device has good sensitivity to microstrains.

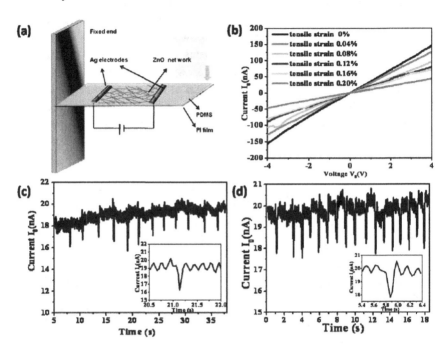

Fig. 2. (a) Schematic illustration of the piezoelectric sensor with ZWN structure and the testing system, by fixing one end to the stage and loading a tensile strain on the other end. (b) The measured I-V curve under ever-increasing tensile strain. (c) The current signal under tensile strain at a frequency of 0.2Hz, inset shows the details of a typical current response of one vibraion. (d) The current signal under tensile strain at a frequency of 0.5Hz, showing good sensitivitises to the microstrains.

With consideration of the extremely small thickness/length of the ZnO layer as compared with the thickness/length of the PI substrate, the strain along the length of the ZWN is approximately calculated by

$$\varepsilon_{zz} = 3\frac{a}{l}\frac{D_{\max}}{l}\left(1 - \frac{z}{l}\right)$$

(1)

where a is the half thickness of the PI substrate, l is the length of the PI substrate from the fixed end to the free end, Dmax is the maximum deformation of the free end of the PI substrate, and z is the distance measured from the fixed end of the PI substrate to the middle of the ZnO layer [13].The maximum ε here is about 0.20%.

A changed I-V characterization after the PI film was stretched increasingly is likely contributed by the following factors. One is the piezoelectric potential generated across the wire; the other is the strain-induced piezoresistance inside the crystal. The change of effective electric tunnels in ZWN structure also should be considered. It has been studied that for a ZnO nanowire with growth direction [0001], once it is bent by an external force, a potential drop is created across the wire, with the stretched surface being positive and the compressed surface being negative [14-16]. Although the free electrons may enter the wires and screen the piezoelectric charges, the piezoelectric potential created by the ions in the crystal cannot be completely neutralized or depleted [17]. In this case, however, the potential was largely consumed by the contacts because of the contact resistance inside the ZWN and between ZWN and the electrodes [18]. Therefore, the effect of piezoelectric potential could be rather limited in this study but still a factor that drives the electrons from the ZnO wires to the electrodes.

Since the ZWN is firmly coated on the PI film, there is a strain imposed on the ZWN due to the deformation of the substrate. Application of strain to a crystal could result in a change in electrical conductivity due to the piezoresistance effect. Strain-induced carrier mobility change and surface modifications have been proved to have clear influence on piezoresistance coefficients [8]. When ZWN were deflected under a tensile strain, the ion distances became larger with the lattice deformations, resulting in carrier mobility decreased because of carrier scattering. The change of crystal resistance remained stable as long as the deformation unchanged. Therefore, piezoresistance effect could be a critical factor to explain the downward current characteristic in Fig. 2(b). To consider the ZWN structure in this work, the contacts between ZnO wires cannot be neglected. When applied a tensile strain, the network was stretched, reducing the contacts of ZnO wires and the active parallel resistances. Hence, the current performed smaller under external tensile strain or tensile vibration.

Typically, the performance of a strain sensor is characterized by a gauge factor (GF), which is the ratio of the normalized change of the electrical conductance of the material to the applied compressive stress [8].The gauge factor can be calculated using the modified formula as shown below:

$$GF = [\Delta I(\varepsilon) / I(0)] / \Delta \varepsilon \qquad (2)$$

186

Fig. 3 shows the highest gauge factor of the fabricated microstrain sensors under 4V bias voltage and tensile strain is above 350, which are relatively higher than the gauge factor of conventional metal strain gauges (about 1-5) [19]. However, further work is needed and accurate electric models should be introduced.

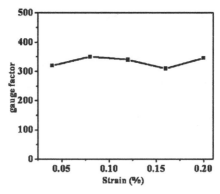

Fig. 3. Gauge factors of the strain sensor under tensile strains

4. Conclusion

In summary, a flexible strain sensor is fabricated with piezoelectric ZnO coated on PI film, by using simple cost-effective hydrothermal method and dispersing process. The I-V characteristics and the current response curves to external vibration with various frequencies proved that the device has good performance to microstrain . When a tensile strain is applied on the ZWN structure, a sharp decreased current curve is received. Both piezoelectric potential and piezoresistance effect contribute to the current changes, but the unique ZWN structure could be a main reason. High gauge factors (~350) demonstrate good sensitivity for tiny static strains and dynamic vibrations. Besides, flexible PI substrate enables the device to work on rough planes and in complex environment. Overall, the fabricated strain sensor has a potential for using in electronic industry and structure monitoring field.

Acknowledgments

This work was supported by the Major Project of International Cooperation and Exchanges (2006DFB51000), NSFC (51172022, 51002008), NSAF (10876001), the Research Fund of Co-construction Program from Beijing Municipal Commission of Education, the Fundamental Research Funds for the Central Universities.

References

1. W. U. Huynh, J. J. Dittmer and A. P. Alivisatos, Science **295** (5564), 2425 (2002).
2. B. Sun, E. Marx and N. C. Greenham, Nano Lett. **3** (7), 961 (2003).
3. Q. Wan, Q. H. Li, Y. J. Chen, T. H. Wang, X. L. He, J. P. Li and C. L. Lin, Appl. Phys. Lett. **84** (18), 3654 (2004).
4. E. R. Leite, I. T. Weber, E. Longo and J. A. Varela, Adv. Mater. **12** (13), 965 (2000).
5. H. M. Hiep, H. Yoshikawa, M. Saito and E. Tamiya, ACS Nano **3** (2), 446 (2009).
6. K. I. Park, S. Xu, Y. Liu, G. T. Hwang, S. J. L. Kang, Z. L. Wang and K. J. Lee, Nano Lett. **10** (12), 4939 (2010).
7. Y. Liu, M. Z. Kauser, M. I. Nathan, P. P. Ruden, S. Dogan, H. Morkoc, S. S. Park and K. Y. Lee, Appl. Phys. Lett. **84** (12), 2112 (2004).
8. R. R. He and P. D.Yang, Nature Nanotech. 1, 42 (2006).
9. J. Liu, P. Fei, J. Zhou, R. Tummala and Z. L. Wang, Appl. Phys. Lett. **92**, 173105 (2008).
10. H. Gullapalli, V. S. M. Vemuru and A. Kumar, Small **6** (15), 1641 (2010).
11. S. N. Cha, J.-S. Seo, S. M. Kim, H. J. Kim, Y. J. Park, S.W. Kim and J. M. Kim, Adv. Mater. **22** (42), 4726 (2010).
12. Z. L. Wang, Adv. Mater.**17**, 889 (2007).
13. J. Zhou, Y. Gu, P. Fei, W. Mai, Y. Gao, R. Yang, G. Bao and Z. L. Wang, Nano Letters **8** (9), 3035 (2008).
14. Z. L. Wang and J. Song, Science **312** (5771), 242 (2006).
15. X. Wang, J. Song, J. Liu and Z. L. Wang, Science **316** (5821), 102 (2007).
16. J. Zhou, P. Fei, Y. Gu, W. Mai, Y. Gao, R. Yang, G. Bao and Z. L. Wang, Nano Let. **8** (11), 3973 (2008).
17. R. Yang, Y. Qin, L. Dai and Z. L. Wang, Nature Nanotech. 4, 34 (2009).
18. J. Zhou, P. Fei and Y. Gao, Nano Lett. **8** (9), 2725 (2008).
19. J. Cao, Q. Wang and H. Dai, Phys. Rev. Lett. **90** (15), 157601 (2003).

STRAIN SENSORS BASED ON SINGLE HIGH-QUALITY ZNO MICROWIRES

ZHIWEI LIU, XIAOQIN YAN[§]

State Key Laboratory for Advanced Metals and Materials, School of Materials Science and Engineering, University of Science and Technology Beijing, Beijing 100083, People's Republic of China

YUE ZHANG

Key Laboratory of New Energy Materials and Technologies, University of Science and Technology Beijing, Beijing 100083, People's Republic of China

Using single ZnO microwires, two strain sensors were fabricated. The sensor I made by a two-ends-bonded ZnO microwire was applied a tensile strain, demonstrating a strain-induced increase in the resistance. The coupled piezoresistance and piezoelectric effects were used to expain the phenomenon. The one-end-bonded ZnO microwire based sensor II was deflected at the free end, showing a linear relationship between the resistance and the bending degree, which is suggested to be associated with the piezoresistance effect. The work exhibits that the ZnO microwires are promising building blocks that can be integrated into the MEMS as a strain sensor for measuring tiny displacement.

1. Introduction

One-dimensional-structured ZnO materials are among the most important multifunctional building blocks for fabricating novel devices in NEMS and MEMS. Due to their unique semiconducting and piezoelectric coupled properties, 1-D structured ZnO materials have shown their promising application in electromechanical fields, such as Nanogenerator[1-5], Piezoelectric Field Effect Transistor[6], strain sensor[7-10], and diode[11,12], and so on. Recently, the research on the micro-/nanoelectromechanical system is growing rapidly, for they have a fast response time, a high sensitivity, and low-power consumption. As we all know, the application of strain to ZnO can cause a change in the resistance owing to either the piezoresistance effect or the piezoelectric effect. Up to now, the nanoscale 1-D ZnO structure based strain sensor has been widely

---

[§] Corresponding author, e-mail: xqyan@mater.ustb.edu.cn

investigated [6,8,13], but the working principles of the strain sensors under different loading modes on the microscale are far from being known.

In this paper, we focused on the longitudinal electromechanical properties of single ZnO microwires under both tensile and bending strains. For sensor I, the electric currents under the positive and negative voltage were both lowered but the magnitudes of the declination were different. The coupled piezoresistance and piezoelectric effects were used to illustrate the new phenomenon.The sensor II displayed an increase in conductance in the bending process, which was explained by piezoresistance effect theory.

2. Experimental details

High-quality ZnO microwires, 5-10μm in diameter and 2-3 millimeter in length, were synthesized on a silicon substrate by the chemical vapor deposition method, The detailed description of growth method and characterizations can be found in my previous work [15].

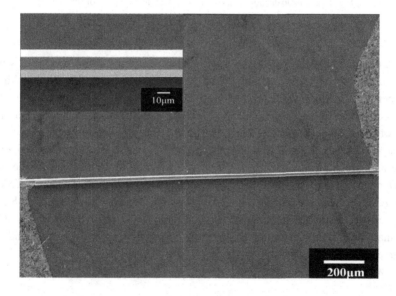

Fig. 1. SEM characterization of high-quality ZnO microwire

To fabricate sensor I, the ZnO microwire was first laid down on a silicon wafer with 20nm SiOx insulation layer, and then each end of the wire was fixed to the substrate using the conductive silver paste; copper wires were also bonded to the ZnO microwire for electrical measurement. A tungsten tip was employed

190

to apply the external force at the middle of the ZnO microwire, producing solely a tensile strain in ZnO microwire. Simultaneously, Keithley 4200 SCS was used to conduct the *in situ* measurement of the electrical properties.

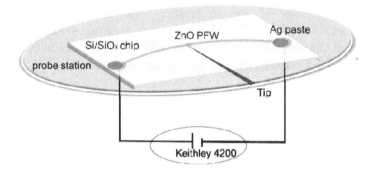

Fig. 2. schematic diagram of electromechanical measurement of sensor I

To create strain sensor II, only one end of ZnO microwire was fixed with the conductive silver paste and linked to the external electric circuit with a copper wire, while the other end was left free. At the same time, the tungsten tip was connected to the keithley 4200 SCS using the other copper wire. When the tip was in contact with the ZnO microwire, the close loop was formed, as illustrated in fig. 3. Every time the microwire was bent for a certain degree, one external sweeping voltage from -2V to 2V was applied. The photographes of ZnO microwire at this time and the corresponding I-V curves were recorded using the camera equipped with the microscope and the Keithley 4200 SCS, respectively. With the increase of bending degree, the procedure was repeated for many times, thus the relationship of the electrical properties and the bending degrees was carried out.

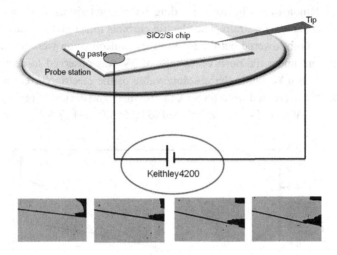

Fig. 3. Schematic diagram of electromechanical measurement of sensor II, and the underlying photographes shows the bending process

3. Results and discussion

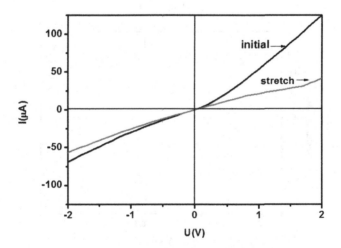

Fig. 4. I-V curves of strain sensor I

Fig. 4 demonstrates the corresponding *I-V* characteristics of sensor I under no strain and the tensile strain. The electric currents under positive and negative voltage were both lowered but the magnitudes of declination were different: the electric current under the positive voltage of 5V was observed to decline from 125 µA to 41.8 µA, while the absolute value of the electric current under the voltage of -5V decreased from 68.8µA to 56.7 µA. The declining rate of electric current under 5V and -5V were calculated to be 66.6% and 17.6%, respectively.

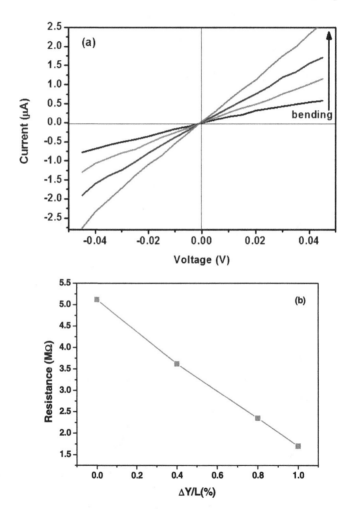

Fig. 5. (a) Current variation with a continuous bending process of the sensor II, and (b) The corresponding relationship of resistance and bending degree.

Repeated experiments were done and the same rule was discovered. For the purpose of understanding the hidden mechanisms, both the piezoresistance and piezoelectric effects were examined. On the basis of the piezoresistance effect, the scattering effect to charge carriers enhanced by the tensile strain is considered to act as a main part that leads to the increase of resistance. In addition, when a tensile stress is applied along the <0001> direction, a piezoelectric potential gradient is produced along the microwire due to the polarization of the ions in the crystal. As a result, the electrons can move easier at one direction while harder at the other direction, thus contributing to the difference between the amplitude of variation in forward and reverse current in IV curves. As a summary, the two effects toke part in the electromechanical properties simultanously in the single ZnO microwire that was imposed by the tensile deformation.

Four sets of IV curves of strain sensor II under different bending degrees were presented in Fig. 5(a), from which the corresponding resistances were calculated. Futhermore, four typical bending degrees of the ZnO microwire were calculated according to the photographes taken by the camera. We simply use the proportion of the deflection of one microwire end to its length to define the bending degree of the ZnO microwire[7]. A clear dependence of conductance on bending strain was found. The ZnO microwire demonstrated a significantly rising conductance in response to the bending strain, suggesting that the strain can enhance the conductivity of ZnO microwire. The relationship between the resistance and the bending degree was depicted in figure 5(b). An almost linear relationship between them was discovered in the entire bending process. The sensitivity of ρ to the mechanical deformation is estimated as $s = d\rho / d\ (\Delta y/L)$ $= 269\Omega \cdot$ cm, indicating that the ZnO microwire could serve as a sensitive microelectromechanical sensors.

For the purpose of proposing the principle of the observed increase of conductance in ZnO microwire, four factors were carefully considered. The first factor is the change in schottky barriers between the ZnO microwire and the tungsten tip. According to the IV characteristics, the typical Ohmic contact was obtained during the measurement. Consequently, the first factor was excluded. The second factor is the change in the defects-related surface properties. Based on the analysis of photoluminescence spectrum of the as-grown ZnO microwire, no oxegen-related defect was found. So this is not the dorminating factor. The third factor is the piezoelectric effect, but as is known to all, it can cause the opposite effect in forward and reverse electric current, which is contrary to what we have observed. The last factor is the piezoresistance effect that is usually caused by change of band gap width as a result of the strained lattice. During the bending process, one side of the microwire was stretched and the other side was compressed. It is generally accepted that a decrease of conduction can be induced by the stretched side while an increase of conductance is caused by the compressed side. As stated by Jun Zhou[8], a strained ZnO most likely has an increased conductivity. Peidong Yang's group quantatively researched on the effect of compressed and stretched strain on silicon nanowire and it was found that the compressed strain played a more impotant role in the change of conductance[14]. In our experiment, the stretched and compressed strain exist simultanously during bending process, but only the increase of conductance was observed due to the combined action of compressed and stretched strains.

4. Conclusions

In this paper, two strain sensors were fabricated and their electromechanical properties were characterized. It was shown that the conductivity of the sensor I was restricted by the tensile deformation, and it was explained by the coupled piezoresistance and piezoelectric effects. The sensor II exhibited a nearly linear relationship in the resistance and the bending degrees and it was demonstrated to be due to the piezoresistance effect. It is shown that the ZnO microwire can act as the potential strain sensor in future MEMS devices

Acknowledgments

This work was supported NSFC (51172022, 50972011), NSAF (10876001), the Research Fund of Co-construction Program from Beijing Municipal Commission of Education, the Fundamental Research Funds for the Central Universities. X. Q. Yan would like to thank the Beijing novel program (2008B19) and the Program for New Century Excellent Talents in University (NCET-09-0219).

References:

1. J. Song, J. Zhou, Z.L. Wang, Nano Lett, 6 (2006) 1656-62.
2. Z.L. Wang, J. Song, Science, 312 (2006) 242.
3. R. Yang, Y. Qin, C. Li, G. Zhu, Z.L. Wang, Nano Lett, 9 (2009) 1201-05.
4. Z.L. Wang, R. Yang, J. Zhou, Y. Qin, C. Xu, Y. Hu, S. Xu, Materials Science and Engineering: R: Reports, (2010).

5. Y. Hu, C. Xu, Y. Zhang, L. Lin, R.L. Snyder, Z.L. Wang, Adv Mater, (2011).

6. X. Wang, X. Wang, J. Zhou, J. Song, J. Liu, N. Xu, Z.L. Wang, Nano Lett, 6 (2006) 2768-72.

7. K.H. Liu, P. Gao, Z. Xu, X.D. Bai, E.G. Wang, Appl Phys Lett, 92 (2008) 213105.

8. J. Zhou, P. Fei, Y. Gao, Y. Gu, J. Liu, G. Bao, Z.L. Wang, Nano Lett, 8 (2008) 2725-30.

9. J. Zhou, Y. Gu, P. Fei, W. Mai, Y. Gao, R. Yang, G. Bao, Z.L. Wang, Nano Lett, 8 (2008) 3035-40.

10. Y. Yang, W. Guo, J. Qi, Y. Zhang, Appl Phys Lett, 97 (2010) 223107.

11. J.H. He, C.L. Hsin, J. Liu, L.J. Chen, Z.L. Wang, Adv Mater, 19 (2007) 781-84.

12. J. Zhou, P. Fei, Y. Gu, W. Mai, Y. Gao, R. Yang, G. Bao, Z.L. Wang, Nano Lett, 8 (2008) 3973-77.

13. Y. Yang, Q. Liao, J. Qi, W. Guo, Y. Zhang, Phys. Chem. Chem. Phys., 12 (2009) 552-55.

14. R. He, P. Yang, Nature nanotechnology, 1 (2006) 42-46.

15. Z. Liu, X. Yan,Z. Lin, Y.Huang, H. Liu, Y.Zhang, Materials Research Bulletin, (2012), doi:10.1016/j.materresbull.2011.12.008

A DESIGN OF RAINBOW SOLAR CELL: AN ORDERLY GRADIENT OF CDS-CDSE SENSITIZED ZNO SOLAR CELL

XIAOYAN HU, YIWEN TANG·*,

Institute of Nano-science and Technology, Central China Normal University, Wuhan, Hubei, 430079, China

A rainbow solar cell configuration based on CdS-CdSe quantum dots (QDs) orderly assembled onto ZnO nanowire (NW) was designed. The rainbow configuration involves alternate cycles of ZnO NW growth and orderly deposition of CdS and CdSe on the different ZnO layer. As a demonstration, in the assembly of ZnO NW bilayer, the presence of CdS shell on the first ZnO NW layer can effectively avoid the fusion of the first ZnO NW layer at the root. Thus the internal surface area of the bilayer assembly is largely enhanced. When the bilayer assemblies were used to fabricate quantum-dot-sensitized solar cells (QD-SSCs), a power conversion efficiency (η) of 0.197% was obtained which was higher than that of conventional ZnO/CdS based QDSSCs. Such a rainbow QD-SSC allows one to couple high electron injection rate of QDs and wide absorption range effectively.

Owing that QDs have efficient charge separation, band gap tunability and higher extinction coefficient than dyes,[1-4] quantum-dot-sensitized solar cells (QD-SSCs) have recently been studied as a promising low-cost alternative compared to conventional dye-sensitized solar cells (D-SSCs). Various QDs, such as CdSe, CdS, PbS, and CdTe have been applied for QDSSCs.[5-8] A common strategy to utilize QDs in solar cells is to couple them with a wide band-gap semiconductor such as TiO_2, ZnO, or SnO_2. In these solar cells, the incident photons firstly excite electrons at ground state in QDs, which generate many electron-hole pairs, and then the charge separation takes place at the interface between QDs and a large band gap semiconductor layer.[9-11] Recently, QDSSCs based on one-dimensional nanomaterials, including nanowires (NWs) and nanotubes (NBs), have been proposed and studied owing to the efficient transport of charge carriers.[12-15]

---

*Corresponding author. Tel.: +86 27 67867947; fax: +86 27 67861185.
E-mail address: ywtang@phy.ccnu.edu.cn

However, QDSSCs based on one-dimensional nanomaterials have shown comparatively lower efficiency than expected. The major challenge of improving the performance of QDSSCs is to enhance the light-harvesting. Some strategies have been employed to improve the light-harvesting. For instance, Lee et al. have fabricated a co-sensitized QDSSC by using two sizes of CdSe QDs.[16] Also QDSSC has been assembled by photosensitizing ZnO NW arrays with CdS and N3 dye.[17] Though ZnO NW arrays have been intensively used in QDSSCs, much lower efficiencies have been achieved than those of porous TiO_2 nanoparticle based QDSSCs. One key challenge of using ZnO NW arrays in QDSSC is that, compared to TiO_2 mesoporous films, the ZnO NWs typical have a low internal surface area, resulting in insufficient QDs assembly and therefore low light harvesting efficiency. Thus, how to optimize the cell configuration and improve the light absorption properties of the electrodes is the main problem in order to improve the performance QDSSCs .To overcome these problems, recently, Dr. Prashant V. Kamat has reported rainbow solar cells which display an ordered assembly of different-sized QDs onto TiO_2 nanotubes. But the idea of rainbow solar cell is rather difficult to realize.[18]

In our previous work, we fabricated ZnO/CdSe core/shell nanowire arrays for efficient solar energy conversion.[19] However, the internal surface area is largely limited for QDs assembling. In this paper, we present an innovative solution to the challenge by synthesizing bilayer-assemblies of ZnO nanowire arrays with orderly gradient of CdS-CdSe decorating. An orderly gradient of CdS-CdSe sensitized ZnO solar cell forms a rainbow configuration. Such a rainbow cell allows one to couple high electron injection rate of QDs and wide absorption range effectively. CdS QDs and CdSe QDs were assembled respectively in an orderly fashion, just as a rainbow displays colors of the visible light spectrum.

A key strategy for preventing the fusion of the first layer at its root during the growth of the second layer of ZnO NW is to coat a CdS shell on the first ZnO layer. A CdS shell coating is used to protect the wires grown in the first layer from widening and fusing in the second ZnO layer growth. This process is schematically shown in Figure 1 and is described as follows. A first layer of ZnO nanowire array was grown on FTO glass substrates (13 Ω/sq) by an ammonia solution method, which was extensively described in previous report. A CdS shell layer was deposited on the first layer of ZnO NW arrays by successive ion layer absorption and reaction (SILAR). The first layer of ZnO NW array was dipped in a 100 mM $CdSO_4$ aqueous solution and subsequenly in 100 mM Na_2S aqueous solution at room temperature. The CdS shell thickness could be controlled digitally by changing the number of reaction cycle. Then, by introducing the obtained ZnO/CdS NW arrays into fresh ZnO growing solution, ZnO/CdS+ZnO NW arrays were obtained. Finally, CdSe QDs were deposited on the bilayer ZnO/CdS+ZnO NW narrays using chemical bath deposition (CBD). The Se source for CdSe deposition was an 80 mM sodium selenossulfate (Na_2SeSO_3) solution, which was prepared by dissolving 0.2 mol Se powder in a 0.5 M Na_2SO_3 solution at 60^OC. Then 80 mM $CdSO_4$ and 160 mM nitrilotriacetic acid trisodium salt (Na_2NTA) were mixed with the 80 mM Na_2SeSO_3 solution. The ZnO/CdS+ZnO NW arrays were immersed into the mixed solution for 6 h.

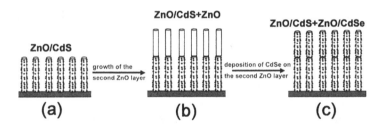

Figure 1. The formation process of the bilayer-assembly of ZnO nanowire arrays with orderly gradient of CdS-CdSe decorating.

Figure 2a shows the scanning electron microscopy (SEM) image of the first-layer of ZnO NW array which was directly grown on FTO. The ZnO NWs are about 6 μm long with the width in the range of 50 nm ~ 100 nm. Prior to the growth of the second layer, a CdS shell was coated on the ZnO NW. Then the substrate was placed into a hydrothermal reaction solution to grow the second layer of ZnO NWs. Figure 2b shows the SEM image of a two-layer assembly of ZnO/CdS+ZnO NW arrays, where an evident boundary between the two layers is observed. The width of the second layer wires was slightly larger than that of the first-layer wires, because as the wires grew vertically they also grew laterally, although at a much slower rate. Figure 2c shows the enlarged SEM image of the first layer of ZnO NW arrays, on which CdS QDs were uniformly deposited. This suggests that the growth of the new-layer wires did not corroded CdS shell on the first layer of ZnO NW and changed the morphology of the previous-layer wire. After CdSe QDs were decorated onto ZnO/CdS+ZnO NW arrays, a typical SEM image (in Figure 2d) of the cross section of sample shows an uniform shell coated on the nanowire surface. In the formation process of the bilayer assemblies, the CdS shell plays an important role. In comparison, the SEM image of the ZnO NW arrays prepared by successive ammonia solution methods of growing ZnO NW array is shown Figure 3. It clearly shows that after the second cycle of growing ZnO NW array, the root of the first layer is fused. Thus, in our formation process of ZnO/CdS+ZnO/CdSe NW arrays, the CdS coating on the sidewall of the first layer ZnO nanowires prevents the aqueous solution from entering the gaps between the wires when the FTO substrate is placed into the hydrothermal reaction solution for growth of the second-layer nanowires, and the reaction solution is only in contact with the wires at their top end. As such, the ZnO wires of the second layer only grows on the top of the first layer, and the first-layer wires are not widened which was confirmed by Figure 2b.

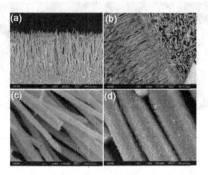

Figure 2. SEM images of (a) the first-layer of ZnO NW array, (b) a two-layer assembly of ZnO/CdS+ZnO NW array. The enlarged SEM images of (c) the first-layer ZnO/CdS NW array, (d) the second-layer of ZnO/CdSe NW array.

Figure 3. SEM image of the ZnO NW array prepared by successive ammonia solution methods of growing ZnO NW array.

The corresponding energy dispersive spectroscopy (EDS) elemental maps of the sample section are recorded in Figure 4. It shows that Zn, O and Cd are

200

present in the whole space. Also Se is richer at the upper side. The spatial distribution of Zn, O, Cd, S, and Se are in accordance with the designed rainbow bilayer structure-orderly assembly of CdS and CdSe onto ZnO NW arrays

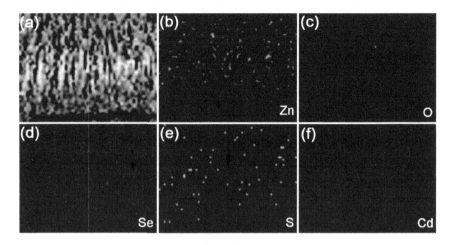

Figure 4. the energy dispersive spectroscopy (EDS) elemental maps of ZnO/CdS+ZnO/CdSe NW array.

The optical absorption spectra of the samples including as-prepared ZnO NW, ZnO/CdS NW, and ZnO/CdS+ZnO/CdSe NW are characterized by UV-vis specterized, as shown in Figure 5. Compared with ZnO NW array, the deposition of CdS moves the absorption onset to a higher wavelength of 500 nm. After assembled with the second layer ZnO/CdSe NW, the absorption edge shifts to the long wavelength almost to 680 nm. These results indicate that the orderly gradient of CdS-CdSe assembled ZnO NW structure enhance the light absorption in almost the entire visible-light region. Moreover, the light absorption intensity is largely enhanced. The back and front side photographs of ZnO/CdS+ZnO/CdSe NW films are shown in Figure 6. Corresponding the ZnO/CdS NW is saffron yellow-colored, and the sample of ZnO/CdS+ZnO/CdSe NW is in dark saffron yellow color. Interestingly, they clearly show that the colour of the back side is paler than that of front side indicating that there is a small quantity of CdSe QDs infiltrated into the first layer.

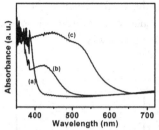

Figure 5. The optical absorption spectra of the samples including as-prepared ZnO NW, ZnO/CdS NW, and ZnO/CdS+ZnO/CdSe NW array.

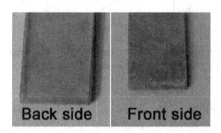

Back side   Front side

Figure 6. The back and front side photographs of ZnO/CdS+ZnO/CdSe NW film.

Such a rainbow cell configuration provides a novel structure for QDSSCs. We assessed the photovoltaic performances of QDSSCs in which ZnO/CdS+ZnO/CdSe NW arrays, ZnO/CdS NW arrays, Pt-coated FTO substrates, and polysulfide redox couple were used as the photoanodes, counter electrode, and electrolyte respectively. The detailed cell fabrication is described as follows: Pt-coated FTO counter electrode was prepared by sputtering. The ZnO electrode was assembled with the counter electrode, and the gap between the two electrodes was controlled by using a 25 μm thick Surlyn film (China). A polysulfide electrolyte composed of 0.5 M Na_2S, 2 M S, 0.2 M KCl in a methanol/water (7:3 by volume) was injected into the gap between the two electrodes.

The photocurrent-voltage (J-V) curves of the assembled solar cells based on ZnO/CdS+ZnO/CdSe NW array and ZnO/CdS NW array were measured under 0.9 sun illumination (Figure 7). The results show that both the short-circuit current density (J_{SC}) and the open-circuit voltage (Voc) for ZnO/CdS+ZnO/CdSe NW array are larger than those for ZnO/CdS. An efficiency (η) of 0.197% was obtained for ZnO/CdS+ZnO/CdSe based solar cell which is much higher than that of ZnO/CdS based solar cell. The above results indicated that there is more injected electron at the ZnO/CdS+ZnO/CdSe photoanode owing to the enhanced light absorption. However, the performance of the devices was still limited by structure issues such as short-circuits through cracks in the devices. The improvement of the devices is under way.

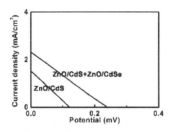

Figure 7. The photocurrent-voltage (J-V) curves of the assembled solar cells based on ZnO/CdS+ZnO/CdSe NW array, ZnO/CdS NW array (tested with a mask of 0.08 cm²).

In summary, we have developed a facile approach to synthesizing an orderly gradient of CdS-CdSe assembled ZnO NW arrays. The approach involves alternate cycles of nanowire growth and the deposition of CdS and CdSe QDs. The internal surface area of the bilayer has been largely enhanced. Such a bilayer assembly has been used to fabricate QDSSCs. It largely increases the capture of incident light by creating an orderly gradient of different QDs. An efficiency of 0.197% was achieved which was much higher than that of ZnO/CdS based QDSSCs. The potential use of such an approach has not been fully explored, and it is believed that by optimizing process parameters the cell performance can be improved.

Acknowledgments

Financially supported by self-determined CCNU basic research and operation research funds for colleges from the China Ministry of Education (CCNU09A02011).

References

1. X. Michalet, F. F. Pinaud, L. A. Bentolila, J. M. Tsay, S. Doose, J. J. Li, G. Sundaresan, A. M. Wu, S. S. Gambhir and S. Weiss, *Science* **538** 28 (2005).
2. W. W. Yu and X. G. Peng, Angew. *Chem. Int. Ed.* **2368**, 41 (2002).
3. W. W. Yu, L. H. Qu, W. Z. Guo and X. G. Peng, *Chem. Mater.* **2854**, 15 (2003).
4. A. J. Nozik, *Inorg.* Chem. **6893**, 44 (2005).

5. I. Yafit, N. Olivia, P. Miles and H. Gary, *J. Phys. Chem. C* **4254**, 113 (2009).
6. H. J. Lee, P. Chen, S. J. Moon, F. Sauvage, K. Sivula, T. Bessho, D. R. Gamelin, P. Comte, S. M. Zakeeruddin, S. I. Seok, M. Gratzel and M. K.Nazeeruddin, *Langmuir.* **7602**, 25 (2009).
7. I. Robel, V. Subramanian, M. Kuno and P. V. Kamat, *J. Am. Chem. Soc.* **2385**, 128 (2006).
8. O. Niitsoo, S. K. Sarkar, C. Pejoux, C. S. Ruhle, D. S. Cahen and G. J. Hodes, *Photochem. Photobiol. A.* **306**, 181 (2006)
9. Y. L. Lee, B. M. Huang and H. T.Chien, *Chem. Mater.* **6903**, 20 (2008).
10. S. Q. Fan, D. Kim, J. J. Kim, D. W. Jung, S. O. Kang and J. Ko, *Electrochem. Commun.* **1337**, 11 (2009).
11. G. Y. Lan, Z. Yang, Y. W. Lin, Z. H. Lin, H. Y. Liao and H. T. Chang, *J. Mater. Chem.* **2349**, 19 (2009).
12. B. Kannan, K. Castelino and A. Majumdar, *Nano Lett.* **1729**, 3 (2003);
13. Y. B. Tang, Z. H. Chen, H. S. Song, C. S. Lee, H. T. Cong, H. M. Cheng, W. J. Zhang, I. Bello and S. T. Lee, *Nano Lett.* **4191**, 8 (2008).
14. L. Hu and G. Chen, *Nano Lett.* **3249**, 7 (2007);
15. O. L. Muskens, J. G. Rivas, R. E. Algra, E. Bakkers and A. Lagendijk, *Nano Lett.* **2638**, 8 (2008).
16. J. Chen, W. Lei and W. Q. Deng, *Nanoscale,* **674**, 3 (2011)
17. X. Song, X. L. Yu, Y. X, J. Sun,T. Ling and X. W. Du. *Semicond. Sci. Technol.* **095014**, 25 (2010).
18. A. Kongkanand, K. Tvrdy, K. Takechi, M. K. Kuno, and P. V. Kamat, *J. Am. Chem . Soc.* **4007**, 130 (2008).
19. Y. W. Tang, X. Y. Hu, M. J. Chen, L. J. Luo, B. H. Li and L. Z. Zhang. *Electrochim. Acta.* **2742**, 54 (2009).

FABRICATION OF GOOD QUALITY N (A-SI) - (NA-SI)-P (MC-SI) TUNNEL JUNCTION FOR TANDEM SOLAR CELLS

MINGJI SHI, YAN ZHANG, LANLI CHEN

Department of Electronics; Nanyang Institute of Technology; Nanyang He Nan 473006
China

We report new results on a tunneling junction for tandem solar cells using a nano-structured amorphous silicon layer (na-Si) as the recombination layer inserted between the n layer and the p layer. Devices were characterized by their dark current-voltage behavior (I-V), activation energy (Ea) and quantum efficiency (QE). The result shows that the tunnel junction with a na-Si insertion layer has higher recombination rates with higher density of defect states of about $2.7 \times 10^{19} cm^{-3}$, lower resistance with activation energy of 22meV. The tunnel junction with a na-Si insertion layer could be easily integrated into the tandem solar cell deposition process.

1. Introduction

Tandem solar cells have attracted extensive interest because of their high conversion efficiency and better stability compared to single junction counterpart [1]. In order to fabricate an efficient tandem solar cell, it is important to design and fabricate good n/p junctions. All photo-generated electrons of the top cell and photo-generated holes of the bottom cell must be recombined at the tunnel junction. If the recombination process does not proceed properly, piled charges will corrupt the electric field inside the cell and the cell performance will be degraded [2]. For tandem solar cell applications, a good n/p junction must allow very high recombination rates, pass current under reverse bias with a low resistance, have negligible optical absorption, and should be easily integrated into the multijunction deposition process [3].

To increase the recombination rate, two methods have been used. One inserts a thin layer of wide band gap foreign material (not a-Si based) such as TiO_x or NbO_x [4, 5]. Wide band gap foreign materials have the advantage that they have high optical transparency and are near-metallic, thus providing ample carriers for recombination. The other inserts a thin layer of heavily doped amorphous silicon (a-Si) [3]. Its benefit is that a-Si is compatible with the plasma enhanced chemical vapor deposition (PECVD) processing of monolithic tandem solar cells [6]. It is desirable to look for a new material as the insertion layer which will have both of the advantages.

In this work, we have made a new attempt to fabricate the tunnel junctions with na-Si insertion layer instead of a-Si layer. Nano-structured amorphous silicon (na-Si: H), also called "nanomorph silicon", has been developed from the SiH_4/H_2 plasma in high plasma power regime of PECVD. It is a diphasic silicon material deposited in the regime at the onset of phase transition from amorphous to nanocrystalline state could gain both the fine photoelectronic properties like a-Si: H and high stability like nanocrystalline silicon. It consists of less than 30% nanocrystallites embedded in amorphous matrix [7, 8]. The nanocrystallites embedded in the amorphous matrix have some good qualities, such as high conductivity, easy to be doped etc [9, 10, 11]. So the na-Si insertion layer can provide more carriers for recombination and is more stable. Our results confirmed that the tunnel junction with na-Si layer is better than its amorphous counterpart because of higher recombination rates and lower resistance and might be a future candidate used in silicon thin film tandem solar cells.

2. Experimental

All the a-Si: H, μc-Si: H and na-Si: H layers were deposited in a three-chambered capacitive coupled radio-frequency (RF of 13.56 MHz) PECVD system. The doping gas ratio of the layers (B_2H_6/ SiH_4) is less than 4.0%; other deposition parameters were listed in Table 1. The thickness of the na-Si insertion layer and the a-Si insertion layer are thinner than 10nm. The a-Si n layer was deposited from SiH_4, H_2, and PH_3 reaction gases and the μc-Si p layer was deposited from SiH_4, H_2, and B_2H_6 reaction gases. The n and p layers were thinner than 50nm. The device structure was stainless steel/ a-Si n / na-Si /μc-Si p /ITO and stainless steel/a-Si n / a-Si /μc-Si p /ITO. Indium tin oxide (ITO) dots (0.07cm^2) were deposited on p-layer as the front electrode by RF magnetron sputtering technique.

Table 1. The deposition parameters of na-Si and a-Si layers.

	Pressure (Pa)	Temperature (°C)	Power density (mW/cm^2)	Hydrogen dilution ratio (H_2/SiH_4)
a-Si	100-300	100-200	50-100	1-10
na-Si	500-1000	50-200	500-1500	100-150

3. Results and discussion

Figure 1 shows the voltage dependence of the resistance (R) at 25°C for the two devices. Plotting the J-V data as R=V/J identifies nonohmic behavior of the devices more clearly than plotting J vs V. Replacing the a-Si layer with na-Si layer lowers the resistance and produces a nearly ohmic behavior. It is clear that the R values of the tunnel junction with the na-Si insertion layer are symmetric about 0V, indicating the device is not operating as n/p diodes. This means the na-Si insertion layer can provide enough recombination centers. If the tunneling junction does not provide enough recombination centers, then a charge will accumulate at this junction, which changes the field distribution near the junction and results in a reverse dipole layer. Furthermore, the junction can be rectifying and generate photovoltage in the reverse direction [5]. Optical transmission of the na-Si layer is almost the same with that of the a-Si layer especially in the long wavelength region. And in both cases the absorption losses in the thin layers are negligible because the thicknesses of the insertion layer are very small (less than 2nm).

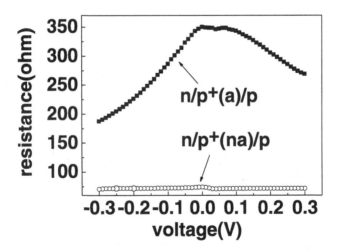

Figure 1. Voltage dependence of resistance (V/J) of the two tunnel junctions at 25°C.

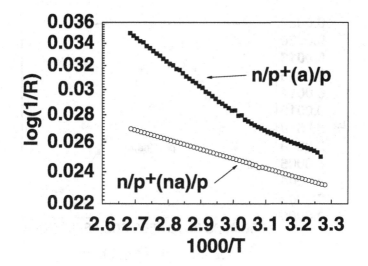

Figure 2. The temperature dependence of conductance of the two tunnel junctions.

Figure 2 shows the temperature dependence of conductance of the two devices. We can derive the Ea of each sample from them. The average Ea of the device with the a-Si insertion layer is 50meV and that of the device with the na-Si insertion layer is 22meV. The Ea indicates the height of the carrier transport barrier crossing the tunnel junction. The Ea value of the device with the na-Si insertion layer is less than KT at room temperature (the thermal excitation energy) indicating there is no barrier to transport across the n/ na-Si /p junction, consistent with the observed lack of rectification.

Figure 3 shows the QE of these same two devices. It is clear that the na-Si layer can reduce the QE more effective than the a-Si layer. We attribute this to the recombination rate is higher in the na-Si layer than in the a-Si layer.

To prove this, a-Si and na-Si thin films were deposited on glass and quartz. The photoconductivity spectra of the films were measured to calculate the density of the band gap defect states of the films. Fig 4 shows the results of the photoconductivity measurement. According to M. Vanecek's theory, the sub-band absorption is caused by the transition from the band gap states to the conduction band (extended states) [12]. The sub-band absorption of our na-Si film is higher than a-Si film means the density of the band gap states of the na-Si film is higher than a-Si film. According to the calculation method proposed by N.

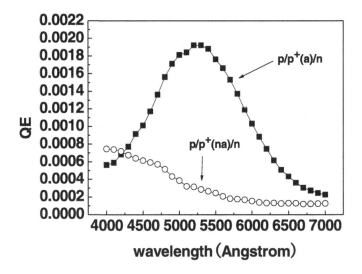

Figure 3. Quantum efficiency at 0 V for the same two devices.

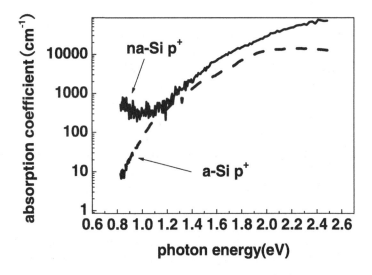

Figure 4. The photoconductivity spectra of the a-Si p⁺ film and the na-Si p⁺ film.

Wyrsch [13], we got the gap states density of our a-Si film is about $1.5 \times 10^{19} \text{cm}^{-3}$ and the gap states density of our na-Si film is about $2.7 \times 10^{19} \text{cm}^{-3}$. This proved na-Si layer can introduce more mid-gap defect states for recombination than a-Si layer. That explained why the insertion of the na-Si layer can greatly improve the quality of the tunnel junction.

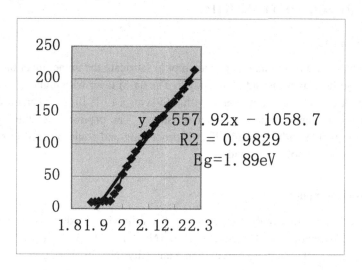

Figure 5. Derive optical band gap of the na-Si p+ layer from transmission spectra.

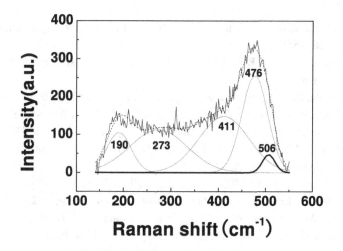

Figure 6. The Raman spectrum of the na-Si p$^+$ film.

The transmission spectra and the Raman shift spectra of the films were also measured. Figure 5 shows the optical band gap of the na-Si film derived from the optical transmission spectrum is about 1.89eV. This proved the film is na-Si: H [7]. The Raman spectrum of the na-Si film is shown in figure 6, we used five Gaussian peaks to simulate the Raman spectrum and the peak near 510cm^{-1} also proved the existence of na-Si: H [14].

4. Conclusions

In this work, we have made a new attempt to fabricate the tunnel junctions with na-Si insertion layer instead of a-Si layer. The na-Si layer with higher gap states density of $2.7 \times 10^{19} \text{cm}^{-3}$ (compared with a-Si layer's $1.5 \times 10^{19} \text{cm}^{-3}$) can introduce higher recombination rate and lower resistance. The deposition of the layer is fully compatible with standard PECVD processing and could be used in silicon thin film tandem solar cells widely.

Acknowledgments

This work was financially supported by the following funds: (1) the Natural Science Foundation of Henan Province (No.092300410182) ;(2) the Key Science and Technology Project of Nan Yang City (No. 2010GG023).

References

1. M. Vukadinovic, F. Smole, M. Topic, et al. *Solar Energy Mater. Solar Cells*. **66**, 361(2001).
2. J. Kwak, S. W. Kwon, K. S. Lim, *J. Non-Crystalline Solids*. **352**, 1847(2006).
3. S. Steven, H. F. Kampas, J. Xi, *Appl. Phys. Lett.* **67**, 813(1995).
4. Y. Sakai, K. Fukuyama, M. Matsumura, et al. *J. Appl. Phys.* **64**, 394(1998).
5. D. Shen R, H. Schropp, R. Chatham, et al. *Appl. Phys. Lett.* **56**, 1871(1990).
6. J. Hou, J. Xi, F. Kampas, et al. *Res. Soc. Symp. Proc.* Vol 336, Boston, Massachusetts. 1994, PP 717-722.
7. Sukti Hazra and Swati Ray. *Jpn. J. Appl. Phys.* **38**, L495 (1999).
8. H. Hao, X. Liao, X. Zeng, et al. *J. Non-crystalline solids*. **352**, 1904(2006).
9. C. Min, W. J. Zhang, T. M. Wang, et al. *Vacuum*, **81**, 126(2006).
10. K. Bhattacharya, D. Das, *Nanotechnology*. **18**, 415704(2007).
11. Y. He, G. Hu, M. Yu, et al. *Phys Rev B*. **59**, 15352(1999).

12. M. Vanecek, J. Kocka, J. Stuchlik, et al. *Solar Energy Mater.* **8**, 411(1983).

13. N. Wyrsch, F. Finger, T. J. Mcmahon, et al. *J. Non-crystalline Solids.* **137**, 347(1991).

14. D. V. Tsu, et al. *Appl. Phys. Lett.* **71**, 1317(1997).

PREPARATION OF GA-DOPED ZNO NANOROD ARRAYS FOR DYE SENSITIZED SOLAR CELLS APPLICATIONS

YING QIU, ZI QIN, YOUSONG GU [†]

State Key Laboratory for Advanced Metals and Materials, School of Materials Science and Engineering, University of Science and Technology Beijing Beijing, 100083, P R China

Ga-doped ZnO nanorods have been prepared directly on fluorine-doped tin oxide (FTO) by a hydrothermal method at 150°C. Crystallization and morphology and effects of Ga doping concentration on efficiency of Dye-sensitized solar cells (DSSCs) were studied by energy dispersive spectroscopy (EDS), x-ray diffraction (XRD), and scanning electron microscopy (SEM). The DSSC using Ga-doped nanorods array exhibited an efficiency of 1.37%, which is much higher than that of DSSC fabricated by undoped ZnO nanorods array. The high energy conversion efficiencies is due to the large carrier concentration in Ga doped ZnO nanorods and high specific surface area of the morphology.

1. Introduction

Due to the advantages of high efficiency, low costs and ease of fabrication, dye-sensitized solar cells (DSSCs) have attracted significant attentions in recent decades [1]. The most promising dye-sensitized system is based on TiO_2 nanocrystallites sensitized by ruthenium dyes [2]. However, in TiO_2-based DSSCs, high efficiencies are limited by the huge number of electrons that recombine with oxidized species from electrolyte or oxidized dye molecules before reaching the collecting electrode [3]. With the similar energy level and high electronic mobility, ZnO is a proper substitution for TiO_2 to fabricate the photo-anode of DSSCs [4-7]. Because of direct pathway for rapid collection of photo-generated electrons, Law et al [8] used ZnO nanorods array to construct DSSCs, which were synthesized in aqueous solution by seeded-growth process. The overall conversion efficiencies of this kind of DSSCs were about 1.2%. Compared with TiO_2-based DSSCs, modifications on ZnO are necessary to improve the conversion efficiency of ZnO based DSSCs. ZnO doped by metal ions, such as Ga[9], Sb[10], and Sn[11] have been reported. Ga doped ZnO

[†] Corresponding anthour. E-mail: yousongu@mater.ustb.edu.cn

showed improved carrier concentration and mobility [12], which could contribute to the enhanced conversion efficiency of DSSCs. But many research works concentrating on preparing Ga doped ZnO nanoparticles by PEMOCVD [13], DC magnetron sputtering, pulsed laser deposition [14] and chemical vapor deposition [15].

In this work, we synthesized Ga doped ZnO nanorods array on FTO substrate by low temperature hydrothermal method, and then Ga doped ZnO was used as photo-anode to fabricate DSSCs. The characterizations indicated that the conversion efficiency has been enhanced by Ga doping.

2. Experimental

2.1. Synthesis and characterization of Ga doped ZnO nanorods arrays

Ga doped ZnO nanorods arrays were prepared by low temperature hydrothermal method using zinc nitrate hexahydrate ($Zn(NO_3)_2 \cdot 6H_2O$), gallium nitrate hydrate ($Ga(NO_3)_3 \cdot xH_2O$), ethylene diamine (EDA, $H_2NCH_2CH_2NH_2$) and sodium hydroxide (NaOH). All the chemicals were purchased commercially and used as received without further purification. $Zn(NO_3)_2$ and $Ga(NO_3)_3$ with a certain mole ratio (Ga/Zn=1%, 3%, 5%, 10%) were dissolved in 75ml de-ionized water. Then 2.5ml of EDA was added into the solution. The PH value of the solution was adjusted to 9~11 by adding NaOH into the former solution while dispersed in ultrasonic apparatus. Then the mixed solution was transferred into autoclave, in which ZnO seed coated FTO glasses were placed and the autoclave was put in furnace and heated at 150°C for 24h. After cooling, the glasses grown with Ga doped ZnO nanomaterials was extracted with tweezers, washed with de-ionized water for several times, and dried in furnace at 60°C for 30min.

Structure, composition and morphology of the samples were studied by X-ray diffraction (XRD, Digaku DMAX-B), Energy dispersive spectroscopy (EDS) and scanning electron microscopy (SEM, SURA TM55). The photoluminescence (PL) was measured by HR800 UV-vis Raman spectrometer (Horiba/Jobin-Yvon).

2.2. Solar cells preparation and characterization

Ga doped ZnO nanorods arrays directly grown on FTOs coated glass substrates (14 ohm/sq, Nippon Sheet Glass, Japan). Then the substrates were immersed in an ethanol solution containing 0.2 mM cis-bis (isothiocyanato) bis (2, 2'-bipyridyl-4, 4'-dicarboxylato)-ruthenium (II) bis-tetrabutylammonium (N719, Dyesol, Australia) at 60°C for 30min for dye loading. The counterelectrode was

Pt coated conductive glass. Sandwich-type DSSCs were fabricated by sealing dye-absorbed electrodes and counterelectrode with plastic (25μm, SX-1170-25, Solaronix). Then liquid electrolyte was injected into the space between two electrodes.

The current-voltage (I-V) characteristics of the DSSCs were measured by an electrochemical interface instruments (Solartron SI 1286/ SI 1260), with a solar simulator (Oriel, 91159A) served as the light source.

3. Results and discussions

SEM images of undoped ZnO sample [Figure 1(a)-(b)] showed array morphology composed of nanorods of about 8μm length and 200nm diameter. With Ga doping, the length of nanorods increased to 12μm with small nanoparticles of 80nm diameter even dispersed on the surface [Figure 1 (c)-(d)]. Due to small nanoparticles, Ga doped ZnO nanorods arrays had higher specific surface area than that of undoped counterparts.

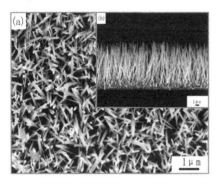

Figure 1. (a)-(b) FE-SEM image of Ga doped ZnO nanorods arrays. (c) - (d) FE-SEM image of Ga doped ZnO nanorods arrays.

XRD patterns of undoped and Ga doped ZnO nanorods arrays [Fig. 2. (a)] revealed that both undoped and doped ZnO were single-crystalline with the wurtzite phase ZnO (JCPDS # 36-1451), there were no characteristic peaks for other impurities such as zinc, gallium and gallium oxide, except ZnO and SnO_2, which is the main component of transparent conductive films on substrate. Small shift in peak position indicated that Ga had entered into the lattice of ZnO because of different atomic radii (1.38Å for Zn and 1.41Å for Ga). EDS results [Fig. 2. (b)] exhibited that the Ga concentrations were identified to be about 0.78%, 1.22%, 1.76% and 2.22% (mole ratio).

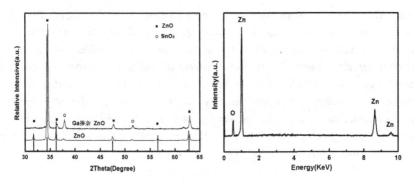

Figure 2. (a) XRD of undoped and 5% Ga doped ZnO nanorods arrays. (b) EDS of Ga doped ZnO nanorods arrays.

Room temperature PL spectra of samples indicated that Ga doped ZnO samples showed an obvious blue shift comparing with those of undoped specimens, as shown in Figure 3. UV emission of undoped ZnO located at about 3.25eV when the light source is He-Cd ion laser (325nm), while the UV emission band of Ga doped ZnO centered at about 3.4eV. The intensity of the UV emission, which reflects recombination of free excitons from ZnO[16-19], decreased and broadened considerably with Ga doping. The wider band gap of gallium oxide could be responsible for the blue shift and its broadening.

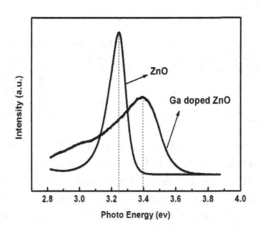

Figure 3. PL spectra of undoped ZnO and Ga doped ZnO nanorods arrays

DSSCs fabricated with Ga doped ZnO nanorods arrays were characterized by measuring the current-voltage behavior while the cells were illuminated

216

under AM1.5 simulated sunlight with a power density of 100mW/cm². Figure 4 exhibits typical current density versus voltage curves of four solar cells using Ga doped ZnO nanorods arrays with different concentrations. Table 1 summarized the open-circuit voltages, short-circuit current densities, fill factors, and overall energy conversion efficiencies for the four samples. It is found that similar open-circuit voltages of approximately 540mV; however, the short-circuit current density increased with Ga doping concentration, which attributed to the improved energy efficiencies from 0.91% to 1.31%. The fill factors were especially low, which limited to 0.37.

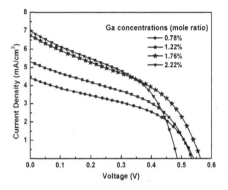

Figure 4. Current-voltage (IV) characteristics for solar cells constructed using different Ga doping content ZnO nanorods arrays.

Table 1. The photovoltaic properties of DSSCs with active area 0.25 cm²

Ga concentration (%)	Voc (V)	Jsc (mA/cm2)	FF (%)	η (%)
0.78	0.54	4.42	0.36	0.91
1.22	0.52	5.30	0.37	1.02
2.22	0.48	6.89	0.40	1.37
1.76	0.57	6.72	0.35	1.34

In a DSSC, open-circuit voltage is determined by the difference between Fermi level of ZnO electrode and I^-/I_3^- redox potential, which accounted for the steady open-circuit voltage of various Ga contents ZnO. The improvement in short-circuit current density is attributed to the increase of charge carriers density and higher specific surface area by Ga doping. Ga-doped ZnO is a kind of N-type doping, which results in the increase of the charge carrier density and the rising of Fermi energy levels. Because of Ga doping, the electrons transport capacity of ZnO is enhanced. And the electrons transport capacity increase leads to the improvement of short-circuit current density. Furthermore, Ga doped ZnO has specific surface area, which contributes to more absorption of dye molecules. The relatively low fill factors were caused by more recombination in Ga doped ZnO than pure ZnO. Defect concentration increased with doping, and more defects created more recombination centers, which result in photo-electrons decreasing before reaching collection layer. Three factors, open-circuit voltage, short-circuit current density and fill factor had comprehensive effects on overall energy conversion efficiencies, and low fill factors lead to decreasing of efficiencies, which brought about limited enhance of efficiencies.

4. Conclusion

In summary, Ga doped ZnO nanowire arrays have been synthesized by low temperature hydrothermal method successfully. The efficiencies of DSSCs fabricated using Ga doping ZnO nanorods arrays increased as the concentration of Ga increased. The optimized efficient of 1.37% at 2.22% Ga concentration is much higher than that of DSSCs made of undoped ZnO nanorods arrays. The main reasons are enhanced charge carrier density and increased specific surface area. However, high defect concentration due to Ga doping also resulted in low fill factors of DSSCs.

Acknowledgments

This work was supported by the Major Project of International Cooperation and Exchanges (2006DFB51000), NSFC(50972009, 51172022), NSAF(10876001), the Research Fund of Co-construction Program from Beijing Municipal Commission of Education, the Fundamental Research Funds for the Central Universities.

References

1. C. Y. Jiang, X. W. Sun, G. Q. Lo, D. L. Kwong, and J. X. Wang, *App. Phys. Let.* **90**, 263501 (2007)
2. M. Grätzel, *Inorganic Chemistry* **44**, 6841 (2005).
3. S. Nakade, Y. Murata, J. Takao, J. Jiu, M. Sakamoto, F. Wang, *J. Am. Chem. Soc.* **126**, 14943(2004).
4. K. Keis, E. Magnuson, H. Lindstrom, S. Lindquist, and A. Hagfeldt, *Sol. Energy Mater. Sol. Cells* **73**, 51 (2002).
5. P. A. Du, H. H. Chen, and Y. C. Lu, *Appl. Phys. Lett.* **89**, 253513 (2006).
6. Y. Wu, H. Yan, M. Huang, B. Messer, J. Song, and P. Yang, *Chem. Eur. J.* **8**, 1260 (2002).

7. Z. G. Chen, Y. W. Tang, L. S. Zhang, and L. J. Luo, *Electrochim. Acta* **51**, 5870 (2006).

8. M. Law, L. E. Greene, J. C. Johnson, R. Saykally, and P. Yang, *Nature materials* **4**, 455 (2005).

9. A. S. Goncalves, M. S. Goes, F. Fabregat-Santiago, T. Moehl, M. R. Davolos, J. Bisquert, S. Yanagida, A. F. Nogueira, and P. R. Bueno, *Electrochim. Acta* **56**, 6503 (2011).

10. N. Ye, J. Qi, Z. Qi, X. Zhang, Y. Yang, J. Liu, and Y. Zhang, *J. Power Sources* **195**, 5806 (2010).

11. S. Y. Bae, C. W. Na, J. H. Kang, and J. Park, *The J. Phys. Chem. B* **109**, 2526 (2005).

12. M. Miyazaki, K. Sato, A. Mitsui, and H. Nishimura, *J. non-cryst. solids* **218**, 323 (1997).

13. V. Khranovskyy, U. Grossner, O. Nilsen, V. Lazorenko, G. V. Lashkarev, B. G. Svensson, and R. Yakimova, *Thin Solid Films* **515**, 472 (2006).

14. M. Yan, H. T. Zhang, E. J. Widjaja, and R. Chang, *J. appl. Phys.* **94**, 5240 (2003).

15. B. M. Ataev, A. M. Bagamadova, A. M. Djabrailov, V. V. Mamedov, R. A. Rabadanov, *Thin solid Films* **260**, 19 (1995).

16. B. D. Yao, Y. F. Chan, and N. Wang, *Appl. Phys. Lett.* **81**, 757 (2002).

17. X. Zhang, Y. Zhang, J. Xu, Z. Wang, X. Chen, D. Yu, P. Zhang, H. Qi, and Y. Tian, *Appl. Phys. Lett.* **87**, 123111 (2005).

18. X. Q. Wei, B. Y. Man, M. Liu, C. S. Xue, H. Z. Zhuang, and C. Yang, *Physica B* **388**, 145 (2007).

19. X. L. Wu, G. G. Siu, C. L. Fu, and H. C. Ong, *Appl. Phys. Lett.* **78**, 2285 (2005).

CONTROLLED SYNTHESIS OF ZNO NANOTETRAPODS AND PERFORMANCE OF ZNO NANOTETRAPODS BASED DYE-SENSITIZED SOLAR CELLS

QI PANG[†], YUEJUN FENG, CHUNJIE LIANG, JUN HE

College of Chemistry and Material, Yulin Normal University, Yulin, Guangxi, 537000, China

LIYA ZHOU

College of Chemistry and Material, Yulin Normal University, Yulin, Guangxi, 537000, China

ZnO nanotetrapods with mean arm diameters of 70, 100, 130, and 165 nm, respectively, were obtained through chemical vapor transfer deposition. The phase and morphology of the ZnO nanotetrapods were investigated using X-ray diffraction spectroscopy and scanning electron microscopy. Photoanode architectures in dye-sensitized solar cells (DSSCs) comprising building blocks of varied arm sizes of ZnO nanotetrapods were introduced. DSSCs based on varied arm sizes of ZnO nanotetrapods were successfully fabricated using the doctor blade technique, and the cell performances were characterized. The highest photoconversion efficiency (η) of 3.21% was achieved using the smallest arm diameter of the ZnO tetrapod film photoanode, which exhibited a shortcircuit current of 9.49 mAcm^{-2}, an opencircuit voltage of 0.649 V, and a fill factor of 0.52 under AM 1.5 irradiation. A variation in η of the DSSCs with the varied arm sizes of ZnO nanotetrapods was observed—η of the DSSCs increased with the decreasing arm diameter of ZnO nanotetrapods.

1. Introduction

Dye-sensitized solar cells (DSSCs), which are inexpensive and easy to fabricate, are one of the several important photovoltaic devices that take advantage of nanostructures to accomplish efficient solar-to-electric power conversion [1]. Grätzel et al. [2] achieved a power conversion efficiency of more than 10% in DSSCs using a TiO$_2$ nanocrystalline thin film sensitized by ruthenium-based dyes. Recently, ZnO is seen to be a promising material for solar cell applications because of its direct wide band gap of 3.37 eV and much higher electron

---

[†] Corresponding author. E-mail address: pqigx@163.com.

mobility (17 cm^2 V^{-1} s^{-1} for ZnO nanowires) compared with that of TiO$_2$ [3]. One-dimensional (1D) nanostructures of ZnO, such as nanowires [4], have been achieved to significantly improve electron transport in the photoanode films by providing a direct conduction pathway for the rapid collection of photogenerated electrons. Martinson et al. introduced ZnO nanotube photoanode DSSCs and achieved 1.6% efficiency [5]. An overall power conversion efficiency of 3.12% was obtained based on ZnO nanorod-nanosheet structured DSSCs [6]. The main method for strengthening the ZnO DSSCs is through the modification of the morphology of the photoanode compared with the conventional 1D structure. The structure of the charge-transporting layer should be optimized to achieve maximum device efficiency. Hsu et al. reported ZnO tetrapods based DSSCs with 1.2% efficiency using of large size tetrapods (about 5 μm in arm length and hundreds of nm in arm diameter)[7]. Yang et al. firstly reported ZnO nanotetrapods based DSSCs with 3,27% efficiency useing well-crystallized ZnO nanotetrapods (about 500 nm in arm length and 40 nm in arm diameter) to construct photoanodes [8]. They found that DSSC based on pure ZnO nanotetrapod photoanode have a reasonable efficiency because the branched nanotetrapods network could effectively transport photoinduced electrons by physically contacting the nanotetrapods. The highest reported photoconversion efficiency for ZnO DSSCs is 6.31%, based on the composite photoanodes of SnO$_2$ nanoparticles/ZnO nanotetrapods [9]. ZnO nanotetrapods, having a branched structure by nature, can be assembled into an interpenetrating connected network. Such a network structure can maintain the characteristics of a vertically aligned nanorod array by significantly reducing the many electron-hopping interjunctions existing in the porous nanotetrapods films. Photoanodes with a highly interpenetrating network can yield high photoconversion efficiency (η) through the enhancement of dye loading. However, reports on the effects of different arm sizes of ZnO nanotetrapods on the DSSCs performances were not reported. In the current paper, ZnO nanotetrapods were synthesized through chemical vapor transfer deposition. Photoanode architectures in DSSCs comprising building blocks of ZnO nanotetrapods with varied arm sizes were introduced. A ZnO film photoanode was prepared via the doctor blade method using ZnO nanotetrapod colloidal pastes. The current study presents the effect of photoanode-based ZnO tetrapod arm size on η of the DSSCs.

2. Experimental procedure

2.1. *Preparation of ZnO nanotetrapods*

Zn foil (0.35 g, 0.25 mm thick) was placed on a small quartz tube with an inner diameter of 6 mm and a length of 60 mm, which was placed at the end of the large quartz tube. The large quartz tube was then inserted into the tube furnace with the closed end at the middle. To remove air in the system, the quartz tube was cleansed for 30 min using N_2 (> 99.9%, 400 standard cubic centimeters per minute (sccm)) poured into the high temperature end of the quartz tube. Then, the N_2 flow rate was kept constant (200 sccm) while the tube furnace was heated to 700 °C at a rate of 25 °C min^{-1}. O_2 flow (45 sccm) saturated with water vapor was injected into the quartz tube at the zone distance 30 cm from the small quartz tube. After being maintained at this temperature for 2 h, the system was allowed to naturally cool down to room temperature. Finally, white products deposited on the inwall of the large quartz tube carefully collected at different temperature zones of 100 °C, 200 °C, 300 °C, and 400 °C were labeled as samples A, B, C, and D, respectively.

2.2. *Preparation of ZnO nanostructure film/fluorine-doped tin oxide (FTO) glass photoanode*

ZnO photoanodes were fabricated using the various synthesized ZnO nanotetrapod arm sizes. The porous ZnO films were deposited on FTO-coated glass substrates with a resistance of 14 Ω per square using the doctor blade technique. Prior to preparation, the FTO glass substrate was ultrasonically rinsed for 15 min in acetone, isopropyl alcohol, and distilled water, successively. ZnO powder (0.4 g) was ground gently using a mortar and pestle and an 8 ml solution of anhydrous terpineol and absolute alcohol (v/v=1:1). Polyethyleneglycol (PEG 2000, 0.05 g) was added dropwise to break up the aggregated ZnO nanotetrapods into a dispersed paste. The paste was spread on the surface of the FTO-conducting substrate through the doctor blade technique using adhesive tape as spacer (thickness limited to 30 μm). Once air-dried, the films were sintered in air for 20 min at 400 °C at a heating rate of 3 °C/min to remove organic materials and for perfect adhesion of the film onto the FTO substrates. The ZnO nanotetrapod photoanode films were obtained after cooling down to room temperature.

2.3. *Assembly of ZnO nanotetrapod DSSCs*

Annealed samples with ZnO nanostructure film/FTO glass photoanode structure used in DSSCs were immersed in a 0.3 mM solution of the dye N-719 (Solaronix) (cis-bis(isothiocyanato) bis (2,2-bipyridyl-4,4-di-carboxylato) ruthenium bis-tetra butylammonium) in absolute ethanol at room temperature for 8 h. Excess unanchored dyes were rinsed off using absolute ethanol, and the films were air dried. Then, the ZnO nanostructure film/FTO glass photoanode structure was completely dried in a vacuum oven at room temperature overnight and then covered with Pt/FTO-glass which served as counter electrode. The counter electrode was made of FTO glass onto which a nanocrystalline Pt decomposed from H_2PtCl_6 was deposited at 450 °C for 20 min. The liquid electrolyte solution used in the solar cells was composed of 0.03 M I_2 (Sigma-Aldrich), 0.05 M LiI, 0.5 M tert-butylpyridine (Sigma-Aldrich), 0.1 M guanidinium thiocyanate, and 0.6 M 1,2-dimethyl-3-propy-imidazolium iodide in acetonitrile. The internal space of the cell was filled with liquid electrolyte through capillary action.

2.4. *Characterization*

The morphology of the ZnO nanotetrapods was studied via scanning electron microscopy (SEM, Philips FEI Quanta 200 FEG). The crystalline structure of the ZnO nanotetrapods was determined via X-ray diffraction spectroscopy (XRD, D/max 2500 v/pc Rigaku). A scanning potentiostat (Electrochemical Station, model LK98BII) was used to scan the potential at a rate of 10 mVs^{-1} and measure the current under the 100 mWcm^{-2} irradiation of a xenon lamp (Oriel, 500 W) with a global air mass (AM) 1.5 for solar spectrum simulation.

3. Results and discussion

3.1. *SEM and XRD characterization*

The characteristic morphology of the ZnO nanoterapods can be observed in the SEM images in Figure 1, which shows the varied arm sizes of the ZnO nanotetrapods collected during different temperature zones. The characteristic morphology of ZnO nanotetrapods is similar to the structure of a tetrahedron. The insets in Figure 1 (a), (b), (c), and (d) are the corresponding enlarged images. Figure 1(a) shows the morphology of the ZnO nanotetrapods in sample A (collected at the temperature zone of 100 °C). These ZnO nanotetrapods have small sizes with a mean arm diameter of 70 nm, which indicates that the distribution of the size is uniform. Figure 1(b) shows the characteristic morphology of sample B (collected at the temperature zone of 200 °C). The average arm diameter of the ZnO tetrapods is 100 nm. Figure 1(c) shows the SEM images of sample C (collected in the temperature zone of 300 °C) with a mean arm diameter of 133 nm. Figure 1(d) shows the SEM images of sample D (collected at the temperature zone of 100 °C) with a mean arm diameter of 165 nm. In general, the tetrapods collected were pure and uniform. The ZnO tetrapods consist of four arms branching from the same center, and the angles between the arms are nearly the same and analogous to the spatial structure of molecular methane.

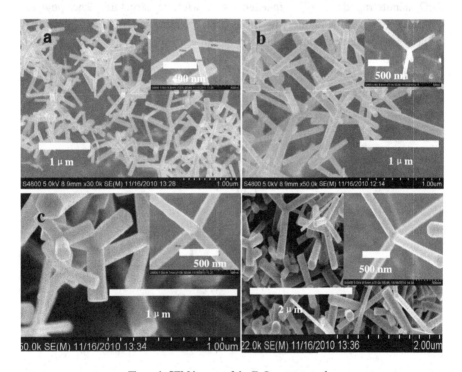

Figure 1. SEM images of the ZnO nanotetrapods.

The growth of the ZnO tetrapods seemed to follow the vapor-solid mechanism. First, Zn was evaporated at a high temperature of 700 °C to provide sufficiently high Zn vapor pressure. This Zn vapor was then transported to the lower temperature zones using a stream of N_2. In the suitable temperature zone of 400 °C, a stream of O_2 was injected, setting off the nucleation and growth of

ZnO tetrapods. The size of the tetrapods was dependent on the local temperature when the Zn vapor pressure was kept constant (the N_2 flow rate was kept constant at 200 sccm). High temperature leads to rapid growth into large tetrapods, whereas low temperature results in slow growth rate. A compromise can be achieved by selecting a suitable O_2 injection temperature that is not too high and not too low. More importantly, the tetrapods formed in higher temperature zones are rapidly transported to lower temperature zones through the N_2 stream, which effectively slows down and terminates their growth.

A typical XRD pattern of the small-sized ZnO tetrapods with a mean arm diameter of 70 nm (sample A) is shown in Figure 2. All diffraction peaks of the ZnO nanotetrapods can be indexed as a wurtzite structure. The powders exhibited a crystalline nature with peaks corresponding to (100), (002), and (101) planes. The preferred orientation corresponding to the (101) plane is observed for the ZnO nanotetrapods.

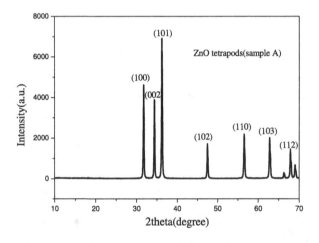

Figure 2. XRD pattern of the small-sized ZnO tetrapods (sample A)

3.2. Cell performances

The current-voltage (J–V) characteristic curves of the four DSSCs are shown in Figure 3 under simulated AM 1.5 light illumination. Curves (a), (b), (c), and (d) are the ZnO tetrapod samples with mean arm diameters of 70 nm, 100 nm, 133 nm, and 165 nm, respectively. The photocurrent density (J) and photovoltage (V) of the cells were measured for the active area of 0.36 cm^2. The highest phtotoconversion efficiency of 3.21 % was achieved with the smallest arm diameter of ZnO tetrapod film photoanodes which exhibited a shortcircuit current (Jsc) of 9.49 mAcm^{-2}, an opencircuit voltage (V$_{oc}$) of 0.649 V, and a fill factor (FF) of 52%. Detailed photovoltaic performance parameters (Jsc, Voc, FF, and η) of the DSSCs for films with varied sizes of ZnO tetrapods are presented in Table 1. η of the DSSCs gradually increased with decreasing arm diameter of the ZnO tetrapods. Therefore, the solar cell performance of the small-sized ZnO nanotetrapod-based DSSCs is better than that of the large-sized ZnO nanotetrapods. This enhancement can be attributed to the electronic transmission efficiency in the small-sized ZnO tetrapod-based DSSCs, which is higher than that in the large-sized ZnO tetrapod-based DSSCs. Moreover, the enhancement is attributed to the increase in the number of dye molecules anchored on the surface of the small-sized ZnO nanotetrapod films, which has larger specific external surface.

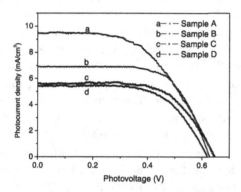

Figure 3. Photocurrent density–photovoltage (J-V) curves of DSSCs based on ZnO nanotetrapods with varied arm diameters.

Table 1. Photovoltaic performance parameters of ZnO nanotetrapod-based DSSCs with varied arm sizes.

sample	Arm diameter (nm)	J$_{sc}$ (mAcm^{-2})	V$_{oc}$ (V)	FF	η (%)
A	70	9.49	0.649	0.52	3.21
B	100	6.92	0.629	0.66	2.91
C	133	5.62	0.647	0.63	2.30
D	165	5.43	0.620	0.61	2.08

The difference in J–V properties may be mainly attributed to the difference in the microstructure of the ZnO films. The small-sized ZnO nanotetrapods are well connected with each neighboring pair in the film-based DSSCs. Although an electron transport pathway in a zigzag pattern is possible, the small-sized ZnO nanotetrapods should exhibit better charge transport than the larger sized ZnO nanotetrapods because with the interconnection of the small-sized ZnO nanotetrapods, a superior porous thin film was developed on the FTO substrate. A porous interpenetrating network was formed, which can provide an efficient pathway to transport the photocurrent. The characteristic symmetrically branched structure of the ZnO nanotetrapods ensures that one of its four arms points to the direction perpendicular to the conductive glass substrate.

4. Conclusion

Varied arm sizes of ZnO tetrapods obtained via chemical vapor transfer deposition were used to fabricate DSSCs using the doctor blade technique. Effects of the photoanode-based ZnO tetrapod arm size on the photoelectric conversion efficiency of the DSSCs were studied. Results indicate that the efficiency of the DSSCs varies with different arm sizes of the ZnO nanotetrapods. η of the DSSCs was reduced with the gradual increase in the ZnO nanotetrapod arm diameter. The highest photoconversion efficiency of 3.21% was achieved using the smallest arm diameter of ZnO tetrapod film photoanode which exhibited a Jsc of 9.49 mAcm^{-2}, a V_{oc} of 0.649 V, and an FF of 52%. The small arm-sized ZnO nanotetrapods, which exhibited better charge transport ability than the large arm-sized ZnO nanotetrapods, resulted in a superior porous thin film on the FTO substrate. This was achieved by the interconnection of the ZnO nanotetrapods which formed a porous interpenetrating network, increasing the number of dye molecules adsorbed on the surface of the small sized-ZnO nanotetrapods. This porous thin film can provide an efficient pathway to transport electrons.

Acknowledgments

The current work was financially supported by the National Science Foundation of China (No. 20863008) and the Natural Science Foundation of Guangxi Province (No. 2011GXNSFA018060).

References

1. Y. Bai, Y. M. Cao, J. Zhang, M. K. Wang, R. Z. Li, P. Wang, S.M. Zakeeruddin and M. Grätzel, Nature Materials. 7,626 (2008).
2. M. Grätzel, *Inorg. Chem.*44, 6841 (2005).
3. Z. Fan, D. Wang, P. C. Chang, W. Y. Tseng and J. G. Lu, *Appl. Phys. Lett.* 85, 5923 (2004).
4. H. Yu, Z. Zhang, M. Han, X. Hao and F. Zhu, *J. Am. Chem. Soc.* 127, 2378 (2005).
5. A. B. F. Martinson, J. W. Elam, J. T. Hupp and M. J. Pellin, *Nano.Lett.* 7, 2183 (2007).
6. J.H. Qiu, M.Guo and X.D.Wang, J.H. Qiu, M.Guo and X.D.Wang, *Appl. Mater. Interfaces.* 3, 2358 (2011).
7. Y. F. Hsu, Y. Y. Xi, C. T. Yip, A. B. Djurisic and W. K. Chan, *Journal of Appl. Phys.* 103, 083114 (2008),.
8. W. Chen, H. F. Zhang, I. M. Hsing and S. H. Yang, *Electrochem. Commu.*, 11, 1057 (2009).
9. W. Chen, Y. C. Qiu, Y.C., Zhong, K. S. Wong and S.H. Yang, *J. Phys. Chem. A.* 114, 3127 (2010).

ELECTROMAGNETIC AND MICROWAVE ABSORPTION PROPERTIES OF CARBONYL IRON/TETRAPOD-SHAPED ZNO NANOSTRUCTURES COMPOSITE COATINGS

HAIBO YU, HUI QIN, YUNHUA HUANG [†]

State Key Laboratory for Advanced Metals and Materials, School of Materials Science and Engineering, University of Science and Technology Beijing Beijing, 100083, P R China

CIP/T-ZnO/EP composite coatings with carbonyl iron powders (CIP) and tetrapod-shaped ZnO (T-ZnO) nanostructures as absorbers, and epoxy resin (EP) as matrix were prepared. The complex permittivity, permeability and microwave absorption properties of the coatings were investigated in the frequency range of 2-18 GHz. The effects of the weight ratio (CIP/T-ZnO/EP), the thickness and the solidification temperature on microwave absorption properties were discussed. When the weight ratio (CIP/T-ZnO/EP), the thickness and the solidification temperature is 28:2:22, 1.8 mm, and 10 ℃, respectively, the optimal wave absorption with the minimum reflection loss (RL) value of -22.38 dB at 15.67 GHz and the bandwidth (RL<-10 dB) of 5.74 GHz was obtained, indicating that the composite coatings may have a promising application in *Ku*-band (12-18 GHz).

1. Introduction

In recent years, along with increasing use of electromagnetic (EM) wave in a wide range of applications like radar systems, silent rooms, mobile devices, wireless network systems, and so on [1-4], problems of the electromagnetic interference (EMI) and electromagnetic compatibility (EMC) are becoming more and more serious, and composite absorbers with high absorption and wide bandwidth are nowadays in great demand [5,6].

To develop excellent microwave absorbing materials with high absorption and wide bandwidth, the composite coatings with carbonyl iron powders (CIP) and tetrapod-shaped ZnO (T-ZnO) nanostructures absorbers, and epoxy resin (EP) matrix were prepared, and the electromagnetic and microwave absorption properties of the composite coatings were investigated in the frequency range of 2-18 GHz.

[†] Corresponding anthour. Email: huangyh@mater.ustb.edu.cn

2. Experimental

T-ZnO nanostructures were synthesized with a tube furnace by the thermal evaporation of Zn powders [7].The CIP powders and EP resins were purchased from market.

The mixture, prepared by uniformly mixing the two absorbers and wax, was poured into a coaxial cylindrical mold with 7.0 mm in outer diameter, 3.04 mm in inner diameter to measure the complex permittivity and permeability with a vector network analyzer HP8722ES.

For the preparation of composite coatings, CIP and T-ZnO nanostructures were well dispersed in ethanol, and then added into the EP matrix. The compound was then sufficiently mixed by constant stirring at 60 ℃ for about 20-30 min, and then coated on an aluminum plate with a size of 180 mm x 180 mm x 5 mm. After the total solidification, the composite coatings were available. The reflection loss versus frequency of the coatings was tested using the arch method. The detailed microwave absorption properties of eleven samples with different weight ratios (CIP/T-ZnO/EP), thicknesses, and solidification temperatures are listed in Table 1.

Table 1. Microwave absorption properties of CIP/T-ZnO/EP composite coatings

Samples	Weight ratio(CIP/T-ZnO/EP)	Thickness (mm)	Solidification Temperature (℃)	Minimum RL value (dB)	Frequency (GHz) (minimum RL)	Bandwidth (GHz) (RL<-10 dB)
1#	28:0:22	1.8	10	-14.51	14.88	5.22
2#	28:1:22	1.8	10	-14.08	14.88	4.68
3#	28:2:22	1.8	10	-22.38	15.67	5.74
4#	28:3:22	1.8	10	-16.92	14.96	5.31
5#	28:4:22	1.8	10	-12.74	16.33	4.26
6#	28:2:22	1.0	20	-3.52	18.00	0
7#	28:2:22	1.4	20	-9.05	18.00	0
8#	28:2:22	1.8	20	-14.18	16.33	4.20
9#	28:2:22	2.2	20	-10.74	12.78	1.90
10#	28:2:22	1.8	30	-8.53	13.76	0
11#	28:2:22	1.8	40	-5.52	13.68	0

3. Results and discussions

3.1. *Morphology of the absorbers and coatings*

Fig. 1 shows the morphology of the two absorbers and the fractured cross-section of the composite coatings. As shown in Fig. 1(a), the as-grown T-ZnO nanostructures have a uniform size with the basal diameter of 200-400 nm and the length of needle bodies of 1-2 μm. From Fig. 1(b), it can be seen that the carbonyl iron powders are of spherical shape with diameters varying from 0.5 to 2 μm. Fig. 1(c) shows a typical SEM image of the fractured cross-section of the coatings under a low magnification, and there are a number of holes which is of great help for the microwave absorption. Fig. 1(d) shows a SEM image of the fractured cross-section under a high magnification, and different regions of the coatings will contribute to the microwave absorption in different ways.

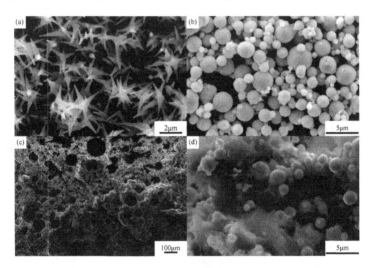

Figure 1. SEM images of (a) T-ZnO nanostructures, (b) carbonyl iron powders and (c) (d) fractured cross-section of 3# coating.

3.2. *Electromagnetic properties*

The frequency dependence of the complex permittivity of the composites with different weight ratios (CIP/T-ZnO/wax) are presented in Fig. 2. Compared to the pure CIP sample, the addition of T-ZnO improves both the real and imaginary parts of permittivity. Fig. 2(a) shows that the real parts of permittivity of 1# - 5# samples keep almost invariable with frequency and the values are about 3.9, 4.3, 4.7, 4.4 and 4.5, respectively. As depicted in Fig. 2(b), the imaginary parts of permittivity increase gradually as the frequency increases due to the increase of the relaxation polarization loss and the electric conductance loss [8], while a dramatic increase of the values from 0.10 at 11.88 GHz to 0.55 at 17.66 GHz occurs for 3# sample with the weight ratio (CIP/T-ZnO/wax) of 28:2:22.

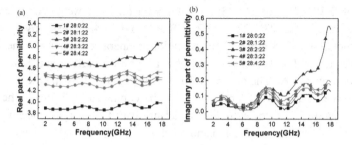

Figure 2. Frequency dependences of (a) the real parts and (b) the imaginary parts of permittivity for composites with different weight ratios (CIP/T-ZnO/wax).

Fig. 3 shows the real and imaginary parts of the complex permeability of 1# - 5# samples. The values of the pure CIP sample (1#) are found to be relatively small, and they are enhanced when CIP were filled with T-ZnO. As can be seen in Fig. 3(a), the real parts of permeability decrease with the increase of frequency in the 2-18 GHz range, and the values for 3# sample are obviously greater than that for other composites in the whole frequency range. Fig. 3(b) shows that, as the frequency increases, the imaginary parts of permeability increase gradually to the maximum at first and then decrease slowly, because of the resonance vibration [9].

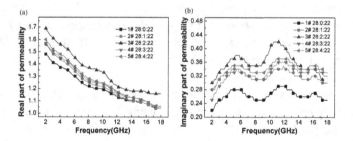

Figure 3. Frequency dependences of (a) the real parts and (b) the imaginary parts of permeability for composites with different weight ratios (CIP/T-ZnO/wax).

3.3. *Microwave absorption properties*

Microwave absorption properties of the coatings with diverse weight ratios (CIP/T-ZnO/EP), thicknesses and solidification temperatures are studied, as shown in Fig. 4.

Fig. 4(a) shows the relationship between reflection loss and frequency in the range of 2-18 GHz for 1# - 5# samples with different weight ratios (CIP/T-ZnO/EP) of 28:0:22, 28:1:22, 28:2:22, 28:3:22 and 28:4:22, respectively, but the same thickness of 1.8 mm and solidification temperature of 10 ℃. The weight concentration of T-ZnO nanostructures has a significant effect on the microwave absorption properties. The minimum RL value decreases first and then rises with the increase of T-ZnO content, which can be explained as follows. The initial decrease is attributed to multipoles resulting from the charge concentration at the needles' tip of T-ZnO nanostructures [10], the interfacial electric polarization [11] and the formation of the conductive network [12]. When the amount of added T-ZnO nanostructures exceeds a limit, impedance mismatch appears and results in weaker wave absorption [13].

Fig. 4(b) shows the frequency dependence of the RL of the samples 6# - 9# with different thicknesses of 1.0, 1.4, 1.8 and 2.2 mm, respectively, but the same weight ratio (CIP/T-ZnO/EP) of 28:2:22 and solidification temperature of 20 ℃. The coating exhibits best microwave absorption behaviors when the thickness is 1.8 mm. Besides, the matching frequency (f_M), at which the minimum RL is available, shifts towards lower frequency with the increase of the thickness, which is in accord with the results reported by L. D. Liu et al. [13].

As shown in Fig. 4(c), the RL spectra of the coatings 3#, 8#, 10# and 11# with different solidification temperatures of 10, 20, 30 and 40 ℃, respectively, but the same weight ratio (CIP/T-ZnO/EP) of 28:2:22 and thickness of 1.8 mm, are presented in the frequency range of 2-18 GHz. The best microwave absorption properties are obtained at the solidification temperature of 10 ℃, and the minimum RL value increases dramatically from -22.38 dB to -5.52 dB with the increase of the temperature, showing worse wave absorption performance, probably due to impedance mismatch.

Fig. 4(d) shows, in the frequency range of 12-18 GHz, the optimal microwave properties of the composite coating that was prepared when the weight ratios (CIP/T-ZnO/EP) is 28:2:22, the thickness is 1.8 mm and the solidification temperature is 10 ℃. The minimum reflection loss value of -22.38 dB was obtained at 15.67 GHz and the bandwidth (RL< -10 dB) is 5.74 GHz in the 12.26-18 GHz range, which shows excellent microwave absorption in *Ku*-band (12-18 GHz).

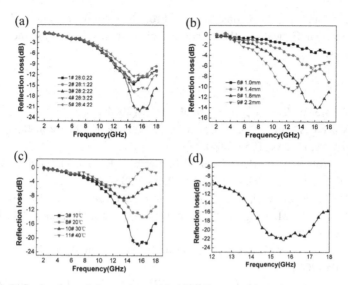

Figure 4. Reflection loss of the coatings with (a) different weight ratios (CIP/T-ZnO/EP), (b) different thicknesses, (c) different solidification temperatures and (d) the weight ratio (CIP/T-ZnO/EP) of 28:2:22, thickness of 1.8 mm and solidification temperature of 10 ℃.

4. Conclusion

CIP/T-ZnO/EP composite coatings with high absorption and wide bandwidth were successfully prepared. Both of the complex permittivity and permeability of the composites can be enhanced by the addition of T-ZnO nanostructures. As T-ZnO nanostructures content in the coatings increases, the minimum RL value first decreases and then goes up. With the increase of the thickness, the matching frequency fm for the minimum RL shifts towards lower frequency. Besides, the minimum RL value increases dramatically with the increase of the solidification temperature, probably because of impedance mismatch. When the weight ratio (CIP/T-ZnO/EP), the thickness and the solidification temperature is 28:2:22, 1.8 mm, and 10 ℃, respectively, the optimal microwave absorption with the minimum RL value of -22.38 dB at 15.67 GHz and the bandwidth (RL<-10 dB) of 5.74 GHz (12.26-18 GHz) was obtained, which indicates that the composite coating may have an important application in *Ku*-band.

Acknowledgments

This work was supported by the Major Project of International Cooperation and Exchanges (2006DFB51000), NSFC (51172022), NSAF (10876001), the

Research Fund of Co-construction Program from Beijing Municipal Commission of Education, the Fundamental Research Funds for the Central Universities.

References

1. V. B. Bregar, *IEEE Trans. Magn.* **40**, 1679 (2004).
2. T. Maeda, S. Sugimoto, T. Kagotani, N. Tezuka and K. Inomata, *J. Magn. Magn. Mater.* **281**, 195 (2004).
3. C. L. Holloway, R. R. Delyser, R.F. German, P. Mckenna and M. Kanda, *IEEE Trans. Electromag. Compat.* **39**, 33 (1997).
4. A. M. Trzynadlowski, *IEEE Trans. Power. Electron.* **21**, 693 (2006).
5. R. C. Che, L. M. Peng, X. F. Duan, Q. Chen and X. L. Liang, *Adv. Mater.* **16**, 401 (2004).
6. R. C. Che, C. Y. Zhi, C. Y. Liang and X. G. Zhou, *Appl. Phys. Lett.* **88**, 033105 (2004).
7. Y. Dai, Y. Zhang, Q. K. Li and C. W. Nan, *Chem. Phys. Lett.* **358**, 83 (2002).
8. Y. B. Feng, T. Qiu and C. Y. Shen, *J. Magn. Magn. Mater.* **318**, 8 (2007).
9. Y. B. Feng, T. Qiu, C. Y. Shen and X. Y. Li, *IEEE Trans. Magn.* **42**, 363 (2006).
10. Z. W. Zhou, L. S. Chu and S. C. Hu, *Mater. Sci. Eng. B* **126**, 93 (2006).
11. Y. J. Chen, M. S. Cao, T. H. Wang and Q. Wan, *Appl. Phys. Lett.* **84**, 3367 (2004).
12. Y. L. Cheng, J. M. Dai, D. J. Wu and Y. P. Sun, *J. Magn. Magn. Mater.* **322**, 97 (2010).
13. L. D. Liu, Y. P.Duan, S. H. Liu, L. Y. Chen and J. B. Guo, *J. Magn. Magn. Mater.* **322**, 1736 (2010).

EFFECT OF CORROSION BY DILUTED HCL SOLUTION ON THE ZNO: AL TEXTURE

MINGJI SHI, PING WANG, LANLI CHEN

Department of Electronics; Nanyang Institute of Technology; Nanyang He Nan 473006 China

High quality textured ZnO: Al electrode can improve the energy conversion efficiency of silicon based thin film solar cells. ZnO: Al films were deposited under 200W. Different textured surfaces were got when etching ZnO: Al films with diluted HCl solutions of 0.5% for different times. The transmission spectrum, square resistance and atomic force microscopy (AFM) images of the samples were measured. The dependence of corrosion time on the resistivity, transmittance and surface texture of the samples were studied. With the increasing of the corrosion time, the resistivity increased, the transmittance decreased, the root-mean-square roughness first increases, then decreases. High quality textured ZnO: Al electrode was obtained when etching the ZnO: Al film deposited under 200W of sputtering power with diluted HCl solution of 0.5%.

1. Introduction

Silicon-based thin film solar cells are favored for less material and low cost, but also confront with the problem of low energy conversion efficiency. The main reason is that the absorption coefficient decreased rapidly with the increase of the incident wavelength. So the absorption length is required to increase significantly to absorb the long wavelength photons. But the thickness of thin film solar cells is small which shorten the transmission distance of photons. Quite a part of the incident photons escaped from the solar cells directly before they were absorbed and converted into photo-generated carriers. This will lead to the waste of long wavelength photons [1]. The light trapping structure could decrease the lost of incident photons by making them being reflected many times in the cells. So the thickness of the solar cells is reduced. Thus, it can improve the efficiency and stability of the cells to achieve the best light trapping structure by controlling the haze of the transparent electrodes, optimizing the parameters of the composite electrodes and adjusting the matching between the transparent electrodes and the cells [2].

In recent years, the textured transparent ZnO: Al films are used more and more widely as the front electrode of thin film solar cells instead of SnO_2 films. So ZnO: Al films become the hotspot of researchers at home and abroad in recent years [3].There are various methods to fabricate ZnO: Al films: magnetron sputtering, ion beam sputtering deposition, vacuum evaporation deposition, sol-gel (Sol-Gel), metal organic chemical vapor deposition (MOCVD) and molecular beam extension (the MBE), etc. Among them, magnetron sputtering is the most popular method. It has the advantage of high deposition rate, low prepare temperature, high film quality and easy to handle [4]. The ZnO: Al films deposited by magnetron sputtering usually need to be textured before they are used.

In 1997, O. Kluth et al [5] proposed the process of forming textured ZnO: Al films by corroding them with diluted HCl solutions. The fabrication process had already been very mature after years of research and practice. It is generally believed that the diluted HCl solution of 0.5% can get the best texture [6-8]. This paper also used diluted HCl solutions to accomplish surface texture. The dependence of the corrosion time on the ZnO: Al surface texture was studied.

2. Experiment

The ZnO: Al films were deposited on glass substrates to get the ZnO: Al/glass samples, using RF magnetron sputtering and using ZnO: Al (Al2O3, 2% wt) as target material. During the deposition process, the target-substrate distance was 8 cm, the Ar flow rate was 40SCCM, the background vacuum was 10^{-4}Pa, the substrate temperature was 250°C, the reaction pressure was 0.8Pa, and the sputtering power was 200 W. The surface texture was realized by corrosion the ZnO: Al film with diluted HCl solution. The dependence of the optical properties and electrical properties of the ZnO: Al film on the deposition parameters was studied. The transmission spectrum of the ZnO: Al/glass samples were measured by grating monochromator (WDM1-3). The ZnO: Al film thicknesses were calculated according to the transmission spectrum. Coplanar Al electrodes were deposited on the film and the square resistance was measured. A layer of Ag film (about 100nm) and a layer of Al film (about 100nm) were deposited on glass in turn to fabricate the Al/Ag/glass substrate. During the deposition process of the Ag film and the Al film, the target-substrate distance was 8 cm, the Ar flow rate was 40SCCM, the background vacuum was 10^{-4}Pa, the substrate temperature was room temperature, the reaction pressure was 0.8Pa, and the sputtering power was 100 W. Then the ZnO: Al films were deposited to get the ZnO: Al/Ag/glass samples. Then the samples were corroded by diluted HCl solution. The AFM was used to scan the sample surface before and after the corrosion process, and the Root-Mean-Square roughness was deduced.

3. Results and discussions

The ZnO: Al/glass samples were dipped into diluted HCl solution of 0.5% to be corroded. The corrosion time (T_C) of sample 1, 2, 3 and 4 was 5s, 10s, 15s and 20s respectively. The thickness (Thk) and square resistance (R_S) of the ZnO: Al films were measured before and after the corrosion process. The dependence of the corrosion rate (R_C) and the square resistance on the corrosion time were calculated, and the interval rate (R_I) was shown in table 1.

Table1. The resistivity of the ZnO: Al films before and after the corrosion process

	Before corrosion			After corrosion			T_C	R_C	R_I
	Thk (nm)	R_S ($\Omega/\square$)	ρ (M$\Omega\cdot$cm)	Thk (nm)	R_S ($\Omega/\square$)	ρ (M$\Omega\cdot$cm)	(s)	(nm/s)	(nm/s)
1	296	26	0.77	278	30	0.83	5	3.6	3.6
2	324	29	0.94	270	45	1.22	10	5.4	7.2
3	351	33	1.16	259	75	1.94	15	6.1	7.6
4	379	40	1.52	249	150	3.73	20	6.5	7.6

It is clear that the interval rate increased fast at the beginning and then kept 7.6 nm/s unchanged after 10s, as shown in table 1. There might be two points to explain why the corrosion rate is small at the beginning. First, the film transformed from amorphous state to crystalline state with the increasing of thickness. The crystallinity of the film was enhanced, the compactness was enhanced and the corrosion resistance was improved. So the corrosion rate at the beginning is small for the corrosion start at the film surface. In addition, when the ZnO: Al samples were just dipped into diluted HCl solutions, the infiltration process would last for a period of time which will make the corrosion rate slow. 5s later, the infiltration process finished and the corrosion rate became steady.

Judging from the square resistance of the film, the square resistance increased with the corrosion time. One reason is that the film thickness decreased with the increasing of the corrosion time. On the other hand, the resistivity (ρ) of the film becomes larger with the increasing of the corrosion time, as shown in table1. The increase of resistivity means that the quality of the film deposited at the beginning is worse than that of the film deposited later.

That might be the result of the film deposited at the beginning was deposited on the amorphous glass, so the crystallinity is fairly bad.

The transmission spectrum of the ZnO: Al film before and after corrosion is shown in figure 1. F0A is the transmission spectra of sample 1 before corrosion whose thickness is 296nm. F5 is the transmission spectrum of sample 1 after corrosion for 5s. F0B is the transmission spectrum of sample 4 before corrosion whose thickness is 379nm. F20 is the transmission spectrum of sample 4 after corrosion for 20s. The thicknesses of sample 2 and sample 3 are between that of sample1 and sample 2. Sample 2 was corroded for 10s and sample 3 was corroded for 15s.

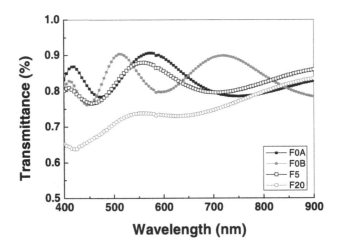

Figure 1. The transmission spectrum of ZnO: Al films before and after corrosion

With the increasing of the corrosion time, the film thickness decreased which means the film's light absorption was reduced. But the transmission spectrum showed that the transmittance decreased with the increasing of the corrosion time. The reason lies in that with the increasing of the corrosion time, the film thickness decreased and the surface began to become coarse, not only transmission but also diffuse transmission took place. As shown in figure 1, the intensity of the interference peak decreased with the increasing of the corrosion time. This means that the surface roughness of the ZnO: Al film increased with the increasing of the corrosion time. While the decreasing of transmittance means that the diffuse transmission increased with the increasing of the surface roughness.

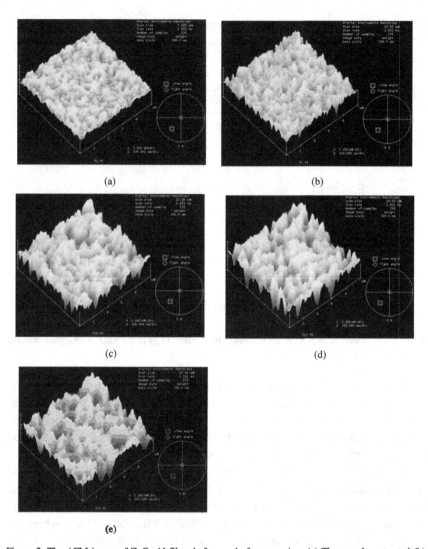

Figure 2. The AFM image of ZnO: Al films before and after corrosion. (a) The sample untreated (b), (c), (d) and (e) the sample treated by the diluted HCl solution of 0.5% for 5s, 10s, 15s and 20s.

To study the changes of the sample surface morphology, AFM was used to scan the sample surface before and after the corrosion process. The AFM image was shown in figure 2. There are bright areas and dark areas in the AFM image. The bright area is raised portion of ZnO: Al surface and the dark area is the sunken portion. After comparison, we can find that the dark spots area increased with the corrosion time which means the size of the pits on the sample surface increased. Through analysis software, we learned that with the increase of the corrosion time the surface roughness increased, the Root-Mean-Square roughness σ_r was shown in figure 3.

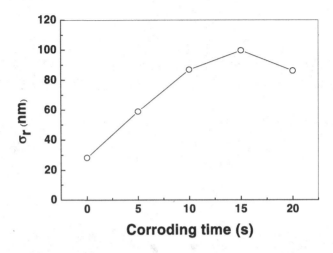

Figure 3. The dependence of Root-Mean-Square roughness (σ_r) of the ZnO: Al/Ag/glass samples on the corrosion time.

This is because the ZnO thin film deposited by magnetron sputtering is polycrystalline material which will show anisotropy when corroded by diluted HCl solution. So the corrosion rate is faster along the direction of some crystalline phase and small pits are formed. With the increasing of corrosion time, the depth and width of the pits would increase. The Root-Mean-Square roughness of the sample decreased from 100 nm to 86 nm when the corrosion time increased from 15 S to 20 S. It is easy to find that the pit bottom of the sample corroded for 15S is sharp and that of the sample corroded for 20S is flat if we compare the AFM images. We speculated that the ZnO: Al film where the corrosion rate is faster has eroded over. But the corrosion rate of the Ag film under the ZnO: Al film is small. So the depth of the pits is limited and only the horizontal corrosion could continue. So the width of the pits is increased and the bottom is flat.

To sum up, high quality textured ZnO: Al electrode was obtained when corrode the ZnO: Al film deposited under 200W of sputtering power with diluted HCl solution of 0.5%.

4. Conclusion

The ZnO: Al samples were fabricated in our magnetron sputtering system under 200W. Different textured surfaces were got when corroding ZnO: Al films with diluted HCl solution of 0.5% for different times. The dependence of the resistivity, transmittance and surface texture of the samples on the corrosion time was studied. The results showed that high quality textured ZnO: Al electrode was obtained when corrode the ZnO: Al film deposited under 200W of sputtering power with diluted HCl solution of 0.5%.

Acknowledgments

The authors wish to acknowledge the many helpful discussions with Haibo Xiao. This work was financially supported by the following funds: (1) the Natural Science Foundation of Henan Province (No.092300410182); (2) the Key Science and Technology Project of Nan Yang City (No. 2010GG023).

References

1. H. H. Li and Q. K. Wang, *Acta Sinica Quantum Optica.* **15**, 380(2009).
2. B. Xu, *Doctoral thesis.* Tianjin: Nankai University, 2005.
3. Q. Lin, X. Gao, Y. Liu, et al. *Chin. J. Vacuum Sci. Technol.* **28**, 575(2008).
4. H. Wang, *Doctoral thesis.* Huazhong University of Science and Technology, 2009.
5. O. Kluth, A. Loffl, S. Wieder, et al. *Proc. 26th PVSC*, Anaheim, CA, 1997.
6. O. Kluth, B. Rech, L. Houben, et al. *Thin Solid Films.* **351**, 247(1999).
7. J. M. Xue, Y. Huang, Q. Xiong, et al. *Acta Armamentarii.* **29**, 504(2008).
8. J. Krc, M. Zeman, O. Kluth, et al. *Thin Solid Films.* **426**, 296(2003).

NANODAMAGE AND NANOFAILURE OF 1D ZNO NANOMATERIALS AND NANODEVICES

PEIFENG LI, YA YANG, YUNHUA HUANG[*], YUE ZHANG

[1]*Department of Materials Physics, University of Science and Technology Beijing, 30 Xueyuan Road, Beijing 100083 China*
[2]*State Key Laboratory for Advanced Metals and Materials, University of Science and Technology Beijing, 30 Xueyuan Road, Beijing 100083 China*

One-dimensional (1D) ZnO nanomaterials include nanowires, nanobelts, and nanorods etc. The extensive applied fields and excellent properties of 1D ZnO nanomaterials can meet the requests of the electronic and electromechanical devices for "smaller, faster and colder", and would be applied in new energy convention, environmental protection, information science and technology, biomedical, security and defense fields. While micro porous, etching pits nanodamage and brittle fracture, dissolving, functional failure nanofailure phenomena of 1D ZnO nanomaterials and nanodevices are observed in some practical working environments like illumination, currents or electric fields, external forces, and some chemical gases or solvents. The more important thing is to discuss the mechanism and reduce or prohibit their generation.

1. Introduction

1D ZnO nanomaterials such as nanowires and nanobelts are being intensively investigated because of their remarkable semiconducting, transverse electrical transport, piezoelectric and thermoelectric, chemical and working stability etc. [1] Based on their excellent properties, many ZnO nanodevices have been fabricated with easily assembly method and low cost, such as photoelectric converters, light-emitting diodes (LEDs), field effect transistors (FETs), field emission devices, nanolasers, nanosensors, nanoelectromechanical system, nanogenerators, and so on. [2-5]

While some phenomena like performance degradation, structure damage, and short life expose under some conditions such as illumination, currents or electric fields, magnetic fields, external pressure or forces, some chemical gases or solvents and so on were observed in the investigation process [6-9] Furthermore, the nanodamage and nanofailure concepts of 1D ZnO nanomaterials and nanodevices have been proposed and attract more and more

---

[*] E-mail: huangyh@mater.ustb.edu.cn

researchers' attention. In order to guarantee the stability, reliability, safety, and long life of the nanodevices in daily life, it is necessary to deeply investigate the security service parameters and failure mechanisms under different applied conditions before the large-scale industrialization of 1D ZnO nanodevices. In addition, the investigation would also guide us to improve and promote the right applications like humidity, chemical gas, biosensors, force/pressure ZnO nanosensors, and to develop some new potential applications like absorbing nanodevices using the absorbing characteristic.

2. Nanodamage and nanofailure of 1D ZnO nanomaterials and nanodevices

In recent years, the structure stability of nanomaterials has attracted many researchers' attention such as the fracture of the carbon nanotubes under the electric field by Liu's group [10], the fracture of the single BN nanotube under the electric field by Bushmaker's group [11], and so on. With the development of the investigation and fabrication of 1D ZnO nanomaterials and nanodevices, performance degradation, structure damage, and short life etc. phenomena also expose in the investigation process under illumination, currents, external pressure or forces, and some chemical gases or chemical solvents. It is necessary to study the nanodamage and nanofailure of ZnO nanomaterials to guarantee the life and safety of the nanodevices. In order to facilitate studying the nanodamage and nanofailure of ZnO nanomaterials, we divide the nanodamage or nanofailure that have been found into mechanical, electric-induced, electromechanical, and chemical nanodamage or nanofailure according to the conditions they working in.

2.1. *Mechanical nanodamage or nanofailure*

Mechanical nanodamage or nanofailure is the structure damage or failure of ZnO nanomaterials under the external forces. Recently, some mechanical nanodamage phenomena of ZnO nanomaterials have been observed by some researchers.

Wen's group reported the brittle fracture of ZnO nanowires with different diameters under external forces and found that the polarity strength of the nanowires increased with the decreasing of the diameter [8]. Zhang's group investigated the mechanical nanodamage of the ZnO nanobelts with different shapes (Fig 1 shows AFM images). They found the nanobelts with triangular cross section would produce mechanical nanodamage easier under the same external force condition than the nanobelts with rectangular cross section. The mechanical nanodamage mechanism may be due to the surface energy of the triangular nanobelt is lower obviously than the rectangular nanobelt.

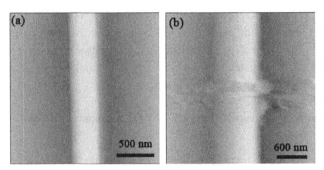

Figure 1. AFM images of ZnO nanobelts under the external forces: the ZnO nanobelt with (a) rectangular cross section and (b) triangular cross section.

2.2. *Electric-induced nanodamage or nanofailure*

Electric-induced nanodamage or nanofailure is the structure damage or failure of nanomaterials under the electron beam or electric field. Till now, few electric-induced nanodamage phenomena of ZnO nanomaterials have been reported.

Zhan's group studied the effect of the convergent electron beam on ZnO nanowires under the TEM [12]. Under the irradiation of 300 KV, micro porous appeared on the smooth surface of the ZnO nanowire before irradiation, but there were no apparent changes in the structure around the micro porous. Wan's group observed the electric-induced fracture in the In-doped ZnO nanobelt under 7.4×10^6 A/cm^2 current density [6]. The researchers reckoned that the In element led the characteristic changes of ZnO from semiconducting to metallic.

Zhang's group observed the obvious fracture phenomenon in the Sb doping ZnO nanobelts under the irradiation of 200 KV convergent electron beam and the damage extent changed more obvious with the increase of time. The fracture phenomenon of the Sb doping ZnO nanobelts was also observed when added voltage using the conductive atomic force tip. The investigation indicated that the Sb element led the structure instability of the ZnO nanobelts.

They also investigated the electric-induced nanodamage (Fig 2 shows AFM images) of the single ZnO nanowire under AFM [7]. They found the nanodamage threshold voltage was about 7~8 V and the nanodamage was more obvious with the increase of time. The nanodamage mechanism may be due to the Joule heat produced under the electric field which led the melt of ZnO nanowires.

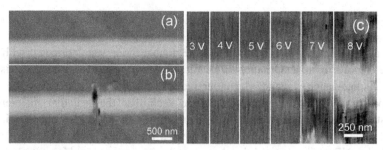

Figure 2. AFM images of a ZnO nanowire (a) before and (b) after electric-induced nanodamage. (c) Electric-induced nanodamage of ZnO nanowires under different applied bias.

2.3. Electromechanical coupling nanodamage

Electromechanical nanodamage or nanofailure is the structural damage or failure of ZnO nanomaterials and nanodevices under the effect of coupling electricity and external forces/pressures. The forms of electromechanical nanodamage and nanofailure can be distinguished by nanomaterials and nanodevices.

Zhang's group studied the electrical and mechanical coupling nanodamage in ZnO nanobelts by using a conductive AFM [13]. The measured damage threshold voltage was found to decrease from 12 to 6 V as the loading forces changed from 20 to 180 nN (Fig 3 shows schematic, AFM images and *F-V* curves). It can be seen that the threshold of voltage of nanodamage changes approximately lineally as the loading forces increase. The mechanism of the decrease in the damage threshold voltages is suggested to be attributed to the strain induced change in electric structures in ZnO.

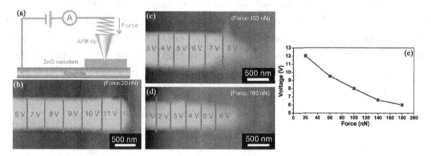

Figure 3. (a) A schematic diagram of nanodamage measurements. AFM images of single ZnO nanobelts under the different applied bias and the different loading forces of (b) 20 nN, (c) 100 nN, and (d) 180 nN. (e) Relationship between the loading forces and nanodamage threshold voltages.

Zhang's group also found the electric-induced nanodamage of ZnO nanodevices using AFM [14]. The single ZnO nanobelt with thickness of about 500 nm was fixed on the 6T film and the double diode was fabricated with the structure of PtIr tip/ZnO nanobelt/6T film/ITO film, and the *I-V* characteristics was studied using a standard conductive AFM with PtIr-coated tips at room temperature. From the plots, it can be seen that the negative differential resistance (NDR) phenomenon was clearly observed in the *I-V* characteristics during the first measurement but disappeared during the second measurement. And the NDR disappeared in the second measurement, which is also not consistent with the electron resonant tunneling, but indicate that the 6T film has broken down after the first measurement.

2.4. *Chemical nanodamage*

Chemical nanodamage or nanofailure is the structure damage or failure of ZnO nanomaterials under the external chemical surroundings. In recent years, some chemical nanodamage of ZnO nanomaterials has been reported by some groups.

Wang's group investigated the chemical nanodamage of the single ZnO nanowire under the acid conditions and alkaline conditions separately [9]. Obvious nanodamage appeared in the surface of the ZnO nanowire after immersed in the acid deionized water, ammonia and NaOH solution for 30 min respectivitly. They also studied the solubility of ZnO nanowires immersed in horse blood serum diluted with 10% aqueous NaOH for different length of time and found that the etching became severe after 3 h and 94% by volume of the ZnO nanowire was dissolved. The study would set the foundation for expanding the application of ZnO nanostructures in bioscience.

Zhang's group investigated the chemical nanodamage (Fig 4 shows the AFM images) of the single ZnO nanobelt in weak alkaline conditions. The etching pits appeared in the surface of the ZnO nanobelt after reaction with ammonia for 15 min and dissolved absolutely after 1 h. The work would provide some meaningful guidance for the applications of ZnO nanodevices in biological detection.

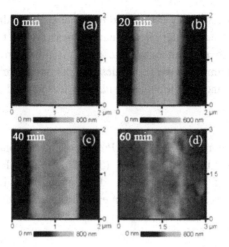

Figure 4. AFM images of ZnO nanowires under the interaction with ammonia for different time: (a) 0 min, (b) 20 min, (c) 40 min, (d) 60 min.

We introduce the typical nanodamage of ZnO nanomaterials that have been observed in the above text briefly. It can be seen that the ZnO nanomaterials and nanodevices would produce micro porous, etching pits etc. nanodamage and brittle fracture, dissolving etc. nanofailure in the poor working conditions like external high pressure, illumination, currents, external forces, and some chemical gases or chemical solvents. And we will continue to concern the nanodamage and nanofailure that produced in different working conditions. Compared with finding the nanodamage and nanofailure, the more important thing is to discuss the mechanism of the nanodamage and nanofailure and reduce or prohibit the nanodamage and nanofailure of the ZnO nanomaterials and nanodevices applied in daily life which are the next work we would do deeply. And the investigation of the nanodamage and nanofailure would provide the designers of nanodevices with useful property parameters to predict the safe working conditions.

3. Conclusions

Mechanical, electric-induced, electromechanical, and chemical nanodamage and nanofailure phenomena of 1D ZnO nanomaterials and nanodevices have been found under corresponding working conditions; and the nanodamage and nanofailure concepts have been proposed which attract many researchers' attention. And it is urgent to investigate the nanodamage and nanofailure of ZnO nanomaterials and nanodevices, and there is still a long way to go to realize the industry of ZnO nanodevices. What need we do urgently is not only to optimize the production processes, but also to find the mechanism of the nanodamage and nanofailure of ZnO nanomaterials and nanodevices for the researchers, and set the safety appliance parameters for the users. By the above investigation, it would be useful to improve the properties of ZnO nanomaterials and guarantee the stability, safety and long life of ZnO nanodevices applied in daily life.

Acknowledgments

This work was supported by the Major Project of International Cooperation and Exchanges (2006DFB51000), NSFC (51172022), NSAF (10876001), the Research Fund of Co-construction Program from Beijing Municipal Commission of Education, the Fundamental Research Funds for the Central Universities.

References

1. Z. L. Wang and J. H. Song, *Science* **312**, 242-246 (2006).
2. X. Wang, J. Zhou, J. Song, J. Liu, N. Xu and Z. L. Wang, *Nano Lett.* **6**, 2768-2772 (2006).
3. X. D. Wang, C. J. Summers and Z. L. Wang, *Nano Lett.* **4**, 423-426 (2004).
4. M. Arnold, Ph. Avouris, Z. W. Pan and Z. L. Wang, *J. Phys. Chem. B* **107**, 659-663 (2003).
5. W. Hughes and Z. L. Wang, *Appl. Phys. Lett.* **82**, 2886-2888 (2003).
6. Q Wan, J Huang, A. X. Lu and T. H Wang, *Appl. Phys. Lett.* **93**, 103109 (2008).
7. Y. Yang, Y. Zhang, J. J. Qi, Q. L. Liao, L. D. Tang and Y. S. Wang, *J. Appl. Phys.* **105**, 084319 (2009).
8. B. M. Wen, J. E. Sader and J. J. Boland, *Phys. Rev. Lett.* **101**, 175502 (2008).
9. J. Zhou, N. S. Xu, and Z. L. Wang, *Adv. Mater.* **18**, 2432-2435 (2006).
10. Y. Wei, P. Liu, K. L. Jiang, L. Liu and S. S. Fan, *Appl. Phys. Lett.* **93**, 023118 (2004).
11. A. W. Bushmaker, V. V. Deshpande, S. Hsieh, M. W. Bockrath and S. B. Cronin, *Nano Lett.* **9**, 607-611 (2009).
12. J. H. Zhan, Y. Bando, Q. J. Hu and D. Golberg, *Appl. Phys. Lett.* **89**, 243111 (2006).
13. Y. Yang, J. J. Qi, Y. S. Gu, W. Guo, and Y. Zhang, *Appl. Phys. Lett.* 2010, **96**, 123103.
14. Y. Yang, J. J. Qi, Q. L. Liao, W. Guo, Y. S. Wang, Y. Zhang, *Appl. Phys. Lett.* 2009, **95**, 123112.

AUTHOR INDEX